D1528173

Elementary Particles
and Symmetries

Documents on Modern Physics

Edited by

ELLIOT W. MONTROLL, *University of Rochester*
GEORGE H. VINEYARD, *Brookhaven National Laboratory*
MAURICE LÉVY, *Université de Paris*
PAUL MATTHEWS, *Imperial College, London*

A. ABRAGAM   L'Effet Mössbauer
S. T. BELYAEV   Collective Excitations in Nuclei
P. G. BERGMANN and A. YASPAN   Physics of Sound in the Sea: Part I Transmission
T. A. BRODY   Symbol-manipulation Techniques for Physics
K. G. BUDDEN   Lectures on Magnetoionic Theory
J. W. CHAMBERLAIN   Motion of Charged Particles in the Earth's Magnetic Field
S. CHAPMAN   Solar Plasma, Geomagnetism, and Aurora
H.-Y. CHIU   Neutrino Astrophysics
A. H. COTTRELL   Theory of Crystal Dislocations
J. DANON   Lectures on the Mössbauer Effect
B. S. DEWITT   Dynamical Theory of Groups and Fields
R. H. DICKE   The Theoretical Significance of Experimental Relativity
H. FESHBACH and F. S. LEVIN   Reaction Dynamics
P. FONG   Statistical Theory of Nuclear Fission
L. GERJUOY, B. YASPAN and J. K. MAJOR   Physics of Sound in the Sea: Parts II and III   Reverberation, and Reflection of Sound from Submarines and Surface Vessels
M. GOURDIN   Lagrangian Formalism and Symmetry Laws
D. HESTENES   Space–Time Algebra
J. G. KIRKWOOD   Dielectrics – Intermolecular Forces – Optical Rotation
J. G. KIRKWOOD   Macromolecules
J. G. KIRKWOOD   Proteins
J. G. KIRKWOOD   Quantum Statistics and Cooperative Phenomena
J. G. KIRKWOOD   Selected Topics in Statistical Mechanics
J. G. KIRKWOOD   Shock and Detonation Waves
J. G. KIRKWOOD   Theory of Liquids
J. G. KIRKWOOD   Theory of Solutions

V. KOURGANOFF   Introduction to the General Theory of Particle Transfer
R. LATTÈS   Methods of Resolution for Selected Boundary Problems in Mathematical Physics
B. W. LEE   Chiral Dynamics
J. LEQUEUX   Structure and Evolution of Galaxies
J. L. LOPES   Lectures on Symmetries
F. E. LOW   Symmetries and Elementary Particles
A. MARTIN and F. CHEUNG   Analyticity Properties and Bounds of the Scattering Amplitudes
P. H. E. MEIJER   Quantum Statistical Mechanics
M. MOSHINSKY   Group Theory and the Many-body Problem
M. MOSHINSKY   The Harmonic Oscillator in Modern Physics: From Atoms to Quarks
M. NIKOLIĆ   Analysis of Scattering and Decay
M. NIKOLIĆ   Kinematics and Multiparticle Systems
H. M. NUSSENZVEIG   Introduction to Quantum Optics
J. R. OPPENHEIMER   Lectures on Electrodynamics
A. B. PIPPARD   The Dynamics of Conduction Electrons
H. REEVES   Stellar Evolution and Nucleosynthesis
L. H. RYDER   Elementary Particles and Symmetries
L. SCHWARTZ   Application of Distributions to the Theory of Elementary Particles in Quantum Mechanics
J. SCHWINGER   Particles and Sources
J. SCHWINGER and D. S. SAXON   Discontinuities in Waveguides
M. TINKHAM   Superconductivity
J. VANIER   Basic Theory of Lasers and Masers
R. WILDT   Physics of Sound in the Sea: Part IV   Acoustic Properties of Wakes

Elementary Particles and Symmetries

LEWIS RYDER
University of Kent at Canterbury

GORDON AND BREACH SCIENCE PUBLISHERS
New York   London   Paris

Copyright © 1975 by
    Gordon and Breach, Science Publishers Ltd.
    42 William IV Street
    London W. C. 2.

Editorial office for the United States of America
    Gordon and Breach, Science Publishers, Inc.
    1 Park Avenue
    New York, N. Y. 10016

Editorial office for France
    Gordon & Breach
    7–9 rue Emile Dubois
    Paris 75014

Library of Congress catalog card number 73-92578. ISBN 0 677 05130 1 (*hardback*) All rights reserved. No part of this book may be reproduced or utilized in any form or by any means, electronic or mechanical, including photocopying, recording, or by any information storage or retrieval system, without permission in writing from the publishers. Printed in Great Britain.

For Mildred and Jessica

Preface

THIS BOOK IS based on lectures given to third year physics undergraduates at the University of Kent, U.K., for the last five years. In it I have tried to present some of the more significant of the recent advances in subnuclear physics in terms which are familiar to undergraduates. Quantum electrodynamics and all reference to quantum field theory are omitted. The Dirac equation is mentioned but knowledge of it is not assumed; and I have tried to keep mathematical details and formalism to a minimum in the hope that anyone familiar with elementary quantum mechanics, special relativity and some nuclear physics would find the book comprehensible.

In general terms, the book describes the symmetry properties of elementary particles and their interactions. Particular attention is paid to isospin and unitary symmetry, which has proved so successful over the last ten years. Chiral symmetry, current algebra, CP violation and the unified gauge theory of weak and electromagnetic interactions are also discussed. These topics seem to me to be among the most interesting and communicable of recent developments. The field of strong interaction dynamics is not dealt with in this book; the interested reader is referred elsewhere.

At the end of each chapter I have compiled a list of references for further reading. These lists are not exhaustive, but it is hoped that they will enable the reader to broaden and deepen his outlook on the subject.

I should like to thank Professor G. Rickayzen and Professor F. Gürsey for their encouraging remarks, and Dr. J. C. Taylor and particularly Professor J. C. Polkinghorne for numerous helpful comments on a draft version of several chapters. I am also grateful to Mr. D. Murray and Mrs. A. Macdonald for preparing the diagrams for the printer, and the secretarial staff of the University of Kent Physics Laboratories and my sister for their help in typing the manuscript. In addition I thank Professors T. D. Lee, J. Steinberger and J. Orear for permission to reproduce diagrams from their papers (Figures 1.3, 7.3 and 8.2 respectively), and Professors A. H. Rosenfeld, Matts Roos and Betty Armstrong for permission to quote extensively from the "Review of Particle Properties" in the table "Elementary Particles and Resonances" at the beginning of this book. The quotation at the end of the book is taken from *Thinkers of the East* by Idries Shah (Jonathan Cape, London).

Finally, and most of all, my thanks go to my wife, without whose constant encouragement and sympathy this book would not exist.

Canterbury   LEWIS RYDER

Contents

Preface ix

Elementary Particles and Resonances xv

1 Introduction 1
   1.1 The lightest elementary particles 1
   1.2 Units and kinematics 1
   1.3 Cross sections and interactions 4
   1.4 Interactions and decay times 9
   1.5 Classification of particles 11
   1.6 Conservation of charge, baryon number and lepton number 12
   1.7 Antiparticles 14

2 Symmetries and Conservation Laws 15
   2.1 Symmetries in classical physics 15
   2.2 Symmetries in quantum physics 18
   2.3 Gauge transformations and other symmetries 22
   2.4 Spin of the charged pion 23
   2.5 Spin of the neutral pion 24
   2.6 Space inversion and parity 24
   2.7 Parity conservation in strong and electromagnetic interactions 29
   2.8 Parity of the photon 30
   2.9 Parity of the charged pion 30
   2.10 Parity of the neutral pion 31
   2.11 Time reversal 31
   2.12 Time reversal invariance in strong and electromagnetic interactions 33
   2.13 Charge conjugation 35
   2.14 C conservation in strong and electromagnetic interactions 36
   2.15 A note on laws of nature and boundary conditions 36
   2.16 Symmetries and groups 37

3 Isospin 41
   3.1 Introduction − a hypothesis 41
   3.2 Evidence for the hypothesis 42
   3.3 Charge symmetry of nuclear forces 43

3.4 Charge independence of nuclear forces 45
3.5 The nucleon and isospin 46
3.6 Isospin transformations 47
3.7 Isospin conservation in strong interactions 52
3.8 Implications of charge independence for pions 52
3.9 Pions and isospin 54
3.10 Raising and lowering operators for the pion 56
3.11 Composition of isospins 58
3.12 Generalised Pauli exclusion principle: the deuteron 59
3.13 Tests for isospin conversion in strong interactions 62
3.14 G parity 63
3.15 Summary 64
Appendix 66

4 **Strangeness** 74
4.1 The discovery of strange particles 74
4.2 The paradox which lead to the introduction of the strangeness quantum number 77
4.3 Isospin of strange particles and the Gell-Mann–Nishijima relation 78
4.4 Parity of the K meson 82
4.5 Parity of Σ 83
4.6 Parity of Ξ 83

5 **Isospin and Strangeness Selection Rules in Weak and Electromagnetic Interactions** 85
5.1 Introduction: leptonic and nonleptonic weak interactions 85
5.2 Strangeness conserving semileptonic weak interactions 86
5.3 Strangeness changing semileptonic weak interactions 89
5.4 Nonleptonic weak interactions 92
5.5 Electromagnetic interactions 94

6 **Electromagnetic Structure of Nucleons** 97
6.1 The difference between electrons and nucleons 97
6.2 Form factors as a description of structure 100
6.3 Proton and neutron form factors 104
6.4 Inelastic electron-proton scattering 107
6.5 Experiments with timelike photons 111

7 **Resonances** 116
7.1 An example: $\Delta(1236)$ 116

7.2 Σ (1385)   119
7.3 Λ (1405) and Λ (1520)   120
7.4 Ξ (1530)   121
7.5 Ω (1672)   121
7.6 ρ meson   122
7.7 ω meson   122
7.8 φ meson   125
7.9 η meson   125
7.10 K^* meson   126
7.11 States which do not exist   128
7.12 Are resonances composite?   128

8 Weak Interactions and Parity Violation 132

8.1 The theta-tau puzzle   132
8.2 The direct observation of parity violation   135
8.3 Parity violation in beta decay   136
8.4 Parity violation and the neutrino   138
8.5 The helicity of the neutrino   141
8.6 $\pi - \mu - e$ decay sequence   143
8.7 Λ nonleptonic decay   143
8.8 CP conservation in beta decay   146

9 K Meson Decays and CP Violation 148

9.1 CPT theorem   148
9.2 Gell-Mann–Pais theory of neutral kaons   150
9.3 K meson interferometry and the $K_1 - K_2$ mass difference   153
9.4 $K_1 - K_2$ mass difference as evidence for the $|\Delta S| = 1$ selection rule   155
9.5 Consequences of CP invariance in K decays   156
9.6 $\Delta I = \tfrac{1}{2}$ rule in $K_1^0 \to 2\pi$ decay   158
9.7 The discovery of CP violation   159
9.8 Analysis of the CP puzzle   159
9.9 Experimental results   164
9.10 The superweak theory   164
9.11 T and CPT in K decays   166
9.12 More on K meson interferometry   166
9.13 CP violation and the arrow of time   168

10 The Conserved Vector Current Theory and Unitary Symmetry   170

10.1 Introduction — the need for a symmetry higher than isospin   170
10.2 Neutron decay and muon decay   171

10.3 The conserved vector current theory 175
10.4 Summary of CVC and steps towards unitary symmetry 179
10.5 Unitary spin supermultiplets: the eightfold way 182
10.6 Unitary spin supermultiplets: the tenfold and onefold ways 185
10.7 Remarks on the meaning of unitary symmetry 187

11 Unitary Symmetry and the Quark Model 190
11.1 The marriage of isospin and strangeness 190
11.2 Building up isospin multiplets 191
11.3 Building up SU_3 multiplets 192
11.4 The quark model 195
11.5 Mass formulae 198
11.6 Mass formulae for electromagnetic mass differences 204
11.7 The significance of the quark model – do quarks exist? 206
11.8 A note on elementarity 206

12 Cabibbo's Theory, Chiral Symmetry and Current Algebra . . 208
12.1 Cabibbo's theory of universality 209
12.2 An interpretation of Cabibbo's theory 211
12.3 Partially conserved axial vector current 213
12.4 Chiral symmetry 216
12.5 Current algebra 220

13 Unified weak and electromagnetic interactions and charm . . 224
13.1 Unified theory of weak and electromagnetic interactions 225
13.2 Charm 231
13.3 Global and local symmetries, and the Higgs mechanism 234
13.4 A note on the new narrow resonances 237
13.5 Concluding remarks 239

Index 243

Elementary Particles and Resonances
(information taken from Review of Particle Properties,
Reviews of Modern Physics, **45**, Supplement, April 1973)

Table of Particles and Resonances

Photon and Leptons

	Spin	L_e	L_μ	Mass (MeV)	Mean Life (sec)	Principal Decay Modes
γ	1	0	0	0 ($<2\times 10^{-21}$)	stable	stable
e^-	½	1	0	0.51	stable	stable
ν_e	½	1	0	0 (<60 eV)	stable	stable
μ^-	½	0	1	105.7	2.2×10^{-6}	$e^-\bar{\nu}_e\nu_\mu$
ν_μ	½	0	1	0 (<1.2 MeV)	stable	stable

Stable* Mesons and Baryons

	B	Spin	I	S	Mass (MeV)	(Mass)² (GeV)²	Mean Life (sec)	Principal Decay Modes	Fraction
$\pi^\pm$	0	0	1	0	139.6	0.019	2.6×10^{-8}	$\mu^\pm\nu_\mu$ $e^\pm\nu_e$	100% 1.2×10^{-4}
π^0	0	0	1	0	135.0	0.018	0.8×10^{-16}	$\gamma\gamma$ γe^+e^-	98.8% 1.2%

	B	Spin	I	S	Mass (MeV)	(Mass)² (GeV)²	Mean Life (sec)	Principal Decay Modes	Fraction
K^+	0	0	½	1	493.7	0.244	1.2×10^{-8}	$\mu^+ \bar{\nu}_\mu$	64%
								$\pi^+ \pi^0$	21%
								$\pi^+ \pi^+ \pi^-$	6%
K^0	0	0	½	1	497.7	0.248			
$\overline{K^0}$	0	0	½	-1	497.7	0.248			
K^-	0	0	½	-1	493.7	0.244	1.2×10^{-8}	$\mu^- \nu_\mu$	64%
								$\pi^- \pi^0$	21%
								$\pi^- \pi^+ \pi^-$	6%
K^0_S	0	0	½				0.9×10^{-10}	$\pi^+ \pi^-$	69%
								$\pi^0 \pi^0$	31%
K^0_L	0	0	½				5.2×10^{-8}	$\pi^0 \pi^0 \pi^0$	21%
								$\pi^+ \pi^- \pi^0$	13%
								$\pi^- \mu^+ \nu_\mu$, $\pi^+ \mu^- \bar{\nu}_\mu$	27%
								$\pi^- e^+ \nu_e$, $\pi^+ e^- \bar{\nu}_e$	39%
η	0	0	0	0	548.8	0.301	2.6 KeV	$\gamma\gamma$	38%
								$\pi^0 \gamma\gamma$	3%
								$3\pi^0$	30%
								$\pi^+ \pi^- \pi^0$	24%
								$\pi^+ \pi^- \gamma$	5%

* Stable: means with respect to strong interaction

	B	Spin	I	S	Mass (MeV)	(Mass)² (GeV)²	Mean Life (sec)	Principal Decay Modes	Fraction
p	1	½	½	0	938.26	0.880	stable	stable	100%
n	1	½	½	0	939.55	0.883	0.92×10^3	$pe^-\nu$	
Λ	1	½	0	−1	1115.6	1.245	2.5×10^{-10}	$p\pi^-$	64%
								$n\pi^0$	36%
								$pe\nu$	8.1×10^{-4}
								$p\mu\nu$	1.6×10^{-4}
Σ^+	1	½	1	−1	1189.4	1.415	0.8×10^{-10}	$p\pi^0$	52%
								$n\pi^+$	48%
								$\Lambda e^+\nu$	2×10^{-5}
Σ^0	1	½	1	−1	1192.5	1.422	$<1.0 \times 10^{-14}$	$\Lambda\gamma$	100%
Σ^-	1	½	1	−1	1197.3	1.434	1.5×10^{-10}	$n\pi^-$	100%
								$ne^-\nu$	1.1×10^{-3}
								$n\mu^-\nu$	0.5×10^{-3}
								$\Lambda e^-\nu$	0.6×10^{-4}
Ξ^0	1	½	½	−2	1314.9	1.729	2.98×10^{-10}	$\Lambda\pi^0$	100%
Ξ^-	1	½	½	−2	1321.3	1.746	1.67×10^{-10}	$\Lambda\pi^-$	100%
Ω^-	1	3/2*	0	−3	1672.5	2.797	1.3×10^{-10}	$\Xi^0\pi^-$ $\Xi^-\pi^0$ ΛK^-	only 28 events seen

Meson States*

	J^P	I^G	S	Mass (MeV)	Width Γ (MeV)	(Mass)^2 (GeV)^2	Mode	Decay Fraction %
$\pi^\pm(140)$	0^-	1^-	0	139.6	0.0	0.0195	See stable particles list	
$\pi^0(135)$				135.0	7.8 eV	0.0182		
$\eta(549)$	0^-	0^+	0	548.8	2.6 keV	0.301	See stable particles list	
ϵ	0^+	0^+	0	≤ 700	≥ 600		$\pi\pi$	
$\rho(770)$	1^-	1^+	0	770	146	0.59	$\pi\pi$	100
				± 5	± 10	± 0.11	e^+e^-	0.004
							$\mu^+\mu^-$	0.007
$\omega(784)$	1^-	0^-	0	783.8	9.8	0.614	$\pi^+\pi^-\pi^0$	90
				± 0.3	± 0.5	± 0.008	$\pi^+\pi^-$	1.3
							$\pi^0\gamma$	9.1
							e^+e^-	0.007
$\eta'(958)$	0^- or (possibly) 2^-	0^+	0	958.1	<2	0.918	$\eta\pi\pi$	72
				± 0.4			$\pi^+\pi^-\gamma$	26
							$\gamma\gamma$	2

* Notation: J = Spin, P = parity, I = isospin, G = G-parity, S = strangeness. Brackets indicate lack of experimental confirmation.

	J^P	I^G	S	Mass (MeV)	Width Γ (MeV)	(Mass)² (GeV)²	Mode	Decay Fraction %
δ(970)	(0⁺)	1⁻	0	~970	~50	0.94	ηπ	
φ(1019)	1⁻	0⁻	0	1019.6 ±0.6	4.2 ±0.2	1.040 ±0.004	K^+K^-	47
							$K_L K_S$	35
							$\pi^+\pi^-\pi^0$	15
							ηγ	3.0
							e^+e^-	0.032
							$\mu^+\mu^-$	0.025
A1(1100)	1⁺	1⁻	0	~1100	200–400	1.21	ρπ	~100
B(1235)	(1⁺)	1⁺	0	1237 ±10	120 ±20	1.53 ±0.12	ωπ	only mode seen
f(1270)	2⁺	0⁺	0	1270 ±5	163 ±15	1.61 ±0.21	ππ	~80
							$2\pi^+2\pi^-$	5 ± 2
							$K\bar{K}$	5 ± 3
D(1285)	(1⁺)	0⁺	0	1286 ±10	30 ±20	1.65 ±0.03	$K\bar{K}\pi$	seen
							ππη	seen
A2(1310)	2⁺	1⁻	0	1310 ±10	100 ±10	1.72 ±0.13	ρπ	72
							ηπ	15
							$K\bar{K}$	5

	J^P	J^G	S	Mass (MeV)	Width Γ (MeV)	(Mass)^2 (GeV)^2	Mode	Decay Fraction %
$E(1420)$	(0^-)	0^+	0	1416 ±10	60 ±20	2.01 ±0.08	$K^*\bar{K} + \bar{K}^*K$ $\eta\pi\pi$	~40 ~60
$f'(1514)$	2^+	0^+	0	1516 ±3	40 ±10	2.29 ±0.06	$K\bar{K}$	only mode seen
$\omega(1675)$		0^-	0	1664 ±13	141 ±17	2.77 ±0.23	$\rho\pi$ 3π	dominant possibly observed
$g(1680)$	3^-	0^-	0	1680 ±20	160 ±30	2.82 ±0.27	2π 4π	~40 ~50
$K^+(494)$ $K^0(498)$	0^-	½	1	493.71 497.71		0.244 0.248	see stable particles list	
$K^*(892)$	1^-	½	±1	891.7 ±0.5	50.1 ±1.1	0.795 ±0.045	$K\pi$	~100
$K_A(1240)$	(1^+)	(½)	±1	1242 ±10	127 ±25	1.54 ±0.16	$K\pi\pi$	only mode seen
$K_A(1280$–$1400)$	(1^+)	½	±1	1280–1400			$K\rho$	seen

	J^P	I^G	S	Mass (MeV)	Width Γ (MeV)	(Mass)² (GeV)²	Decay Mode	Decay Fraction %
$K_N(1420)$	(2^+)	½	±1	1421 ±5	100 ±10	2.02 ±0.14	$K\pi$ $K^*\pi$ $K\rho$ $K\omega$	55 29 9 4
$L(1770)$	(2^-)	½	1	1765 ±10	140 ±50	3.11 ±0.025	$K\pi\pi$	dominant

Baryon States

	J^P	I^G	S	Mass (MeV)	Width Γ (MeV)	(Mass)² (GeV)²	Decay Mode	Decay Fraction %
p n	½⁺	½	0	938.3 939.6		0.880 0.883	see stable particles list	
$N'(1470)$	½⁺	½	0	~1470	165–300	2.16 ±0.41	$N\pi$ $N\pi\pi$	60 40
$N'(1520)$	3/2⁻	½	0	1510–1540	105–150	2.31 ±0.18	$N\pi$ $N\pi\pi$	50 ~50
$N'(1535)$	½⁻	½	0	1500–1600	50–160	2.36 ±0.18	$N\pi$ $N\eta$ $N\pi\pi$	35 55 ~10
$N'(1670)$	5/2⁻	½	0	1670–1685	115–175	2.79 ±0.24	$N\pi$ $N\pi\pi$	40 60

	J^P	I^G	S	Mass (MeV)	Width Γ (MeV)	(Mass)^2 (GeV)^2	Decay Mode	Decay Fraction %
$N'(1688)$	$5/2^+$	$1/2$	0	1680–1690	105–180	2.85 ±0.21	$N\pi$ $N\pi\pi$	60 40
$N''(1700)$	$1/2^-$	$1/2$	0	1665–1765	100–300	2.89 ±0.42	$N\pi$ $N\pi\pi$	60 ~20
$N''(1780)$	$1/2^+$	$1/2$	0	1650–1860	50–350	3.17 ±0.51	$N\pi$ $N\eta$ $N\pi$ ΛK	10–20 25 ~5
$N(1860)$	$3/2^+$	$1/2$	0	1770–1860	180–330	3.46 ±0.57	$N\pi$ $N\pi\pi$	25
$N(2190)$	$7/2^-$	$1/2$	0	2000–2260	270–325	4.80 ±0.67	$N\pi$ $N\pi\pi$	15
$N(2220)$	$(9/2^+)$	$1/2$	0	2200–2245	260–330	4.93 ±0.65	$N\pi$	
$N(2650)$	$?^-$	$1/2$	0	~2650	~360	7.02 ±0.95	$N\pi$	
$N(3030)$	$?$	$1/2$	0	~3030	~400	9.18 ±1.21	$N\pi$	
$\Delta(1236)$	$3/2^+$	$3/2$	0(++)	1230–1236	110–122	1.53 ±0.14	$N\pi$ $N\gamma$	99.4 ~0.6
$\Delta(1650)$	$1/2^-$	$3/2$	0	1615–1695	130–200	2.72 ±0.28	$N\pi$ $N\pi\pi$	28 72
$\Delta(1670)$	$3/2^-$	$3/2$	0	1650–1720	175–300	2.79 ±0.40	$N\pi$ $N\pi\pi$	15
$\Delta(1890)$	$5/2^+$	$3/2$	0	1840–1920	200–350	3.57 ±0.49	$N\pi$ $N\pi\pi$	17

	J^P	J^G	S	Mass (MeV)	Width Γ (MeV)	(Mass)² (GeV)²	Mode	Decay Fraction %
Δ(1910)	½⁺	3/2	0	1780–1935	200–340	3.65 ±0.52	$N\pi$ $N\pi\pi$	25
Δ(1950)	7/2⁺	3/2	0	1930–1980	170–270	3.80 ±0.44	$N\pi$ $N\pi\pi$	45
Δ(2420)	(11/2⁺)	3/2	0	2320–2450	270–350	5.86 ±0.75	$N\pi$ $N\pi\pi$	11 >20
Δ(2850)	?⁺	(3/2)	0	~2850	~400	8.12 ±1.14	$N\pi$ $N\pi\pi$	
Δ(3230)	?	(3/2)	0	~3230	~440	10.4 ±1.4	$N\pi$ $N\pi\pi$	
Λ	½⁺	0	−1	1115.6			see stable particles list	
Λ(1405)	½⁻	0	−1	1405 ±5	40 ±10	1.97 ±0.06	$\Sigma\pi$	100
Λ'(1520)	3/2⁻	0	−1	1518 ±2	16 ±2	2.30 ±0.02	$N\bar{K}$ $\Sigma\pi$ $\Lambda\pi\pi$ $\Sigma\pi\pi$	45 ± 1 41 ± 1 10 1
Λ''(1670)	½⁻	0	−1	~1670	15–38	2.79 ±0.04	$N\bar{K}$ $\Lambda\eta$ $\Sigma\pi$	15–35 15–25 30–50
Λ''(1690)	3/2⁻	0	−1	~1690	27–85	2.86 ±0.09	$N\bar{K}$ $\Sigma\pi$ $\Lambda\pi\pi$ $\Sigma\pi\pi$	20–30 40–70 <25 <25

	J^P	I^G	S	Mass (MeV)	Width Γ (MeV)	(Mass)² (GeV)²	Mode	Decay Fraction %
Λ′(1815)	5/2⁺	0	−1	1820 ±5	64–104	3.30 ±0.15	$N\bar{K}$ $\Sigma\pi$ $\Sigma(1385)\pi$	61 11 15–20
Λ′(1830)	5/2⁻	0	−1	1810–1840	60–140	3.33 ±0.19	$N\bar{K}$ $\Sigma\pi$	~10 20–60
Λ(2100)	7/2⁻	0	−1	~2100	60–140	4.41 ±0.22	$N\bar{K}$ $\Sigma\pi$	25 ~5
Λ(2350)	?	0	−1	~2350	140–324	5.52 ±0.55	$N\bar{K}$	
Λ(2585)	?	0	−1	~2585	~300	6.66 ±0.77	$N\bar{K}$	
Σ	1/2⁺	1	−1	(+) 1189.4 (0) 1192.5 (−) 1197.3		1.41 1.42 1.43	see stable particles list	
Σ′(1385)	3/2⁺	1	−1	(+) 1383 ± 1 (−) 1386 ± 2	(+) 34 ± 2 (−) 36 ± 6	1.92 ±0.05	$\Lambda\pi$ $\Sigma\pi$	90 10
Σ′(1670)	3/2⁻	1	−1	~1670	35–36	2.79 ±0.08	$N\bar{K}$ $\Sigma\pi$ $\Lambda\pi$	~8 ~40 ~12
Σ″(1750)	1/2⁻	1	−1	1700–1790	50–100	3.05 ±0.13	$N\bar{K}$ $\Lambda\pi$ $\Sigma\eta$	seen seen seen

	J^P	I^G	S	Mass (MeV)	Width Γ (MeV)	(Mass)² (GeV)²	Mode	Decay Fraction %
Σ(1765)	5/2⁻	1	−1	1765 ±5	~120	3.12 ±0.21	$N\bar{K}$	~41
							$\Lambda\pi$	~13
							$\Lambda(1520)\pi$	~15
							$\Sigma(1385)\pi$	~10
							$\Sigma\pi$	~1
Σ(1915)	5/2⁺	1	−1	1900–1930	50–100	3.67 ±0.14	$N\bar{K}$	~14
							$\Lambda\pi$	~6
							$\Sigma\pi$	~6
Σ″(1940)	(3/2⁻)	1	−1	~1940	~220	3.77 ±0.43	$N\bar{K}$	seen
							$\Lambda\pi$	seen
							$\Sigma\pi$	
Σ(2030)	7/2⁺	1	−1	~2030	100 −170	4.12 ±0.27	$N\bar{K}$	~20
							$\Lambda\pi$	~20
							$\Sigma\pi$	~4
							ΞK	<2
Σ(2250)	?	1	−1	~2250	100–230	5.06 ±0.37	$N\bar{K}$	
							$\Sigma\pi$	
							$\Lambda\pi$	
Σ(2455)	?	1	−1	~2455	~120	6.03 ±0.29	$N\bar{K}$	
Σ(2620)	?	1	−1	~2620	~175	6.86 ±0.46	$N\bar{K}$	

	J^P	I^G	S	Mass (MeV)	Width Γ (MeV)	(Mass)² (GeV)²	Decay Mode	Fraction %
Ξ	½(+)	½	−2	(0) 1314.9 (−) 1321.3		1.73 1.75	see stable particles list	
Ξ(1530)	3/2(+)	½	−2	(0) 1531.6 ± 0.4 (−) 1535.0 ± 0.6	(0) 9.1 ± 0.5 (−) 12.9 ± 4.1	2.34 ±0.01	Ξπ	100
Ξ(1820)	?	½	−2	1795–1870	12–99	3.31 ±0.10	$\Lambda \bar{K}$ Ξπ Ξ(1530)π $\Sigma \bar{K}$	
Ξ(1940)	?	½	−2	1894–1961	42–140	3.72 ±0.18	Ξπ Ξ(1530)π	
Ω⁻	(3/2⁺)	0	−3	1672.5		2.80	see stable particles list	

CHAPTER 1

Introduction

1.1 THE LIGHTEST ELEMENTARY PARTICLES

THE MOST well-known elementary particles are the electron e^-, proton p and neutron n, which were discovered in the early studies of atoms. The photon γ, the quantum of electromagnetic radiation has been known since the photoelectric effect was discovered; and the existence of the neutrino ν was deduced by studying neutron beta decay $n \to p + e^- + \bar{\nu}$ (in which an antineutrino is emitted). The neutrino like the photon, has zero rest-mass. In 1933 these particles, together with the positron e^+ and antineutrino $\bar{\nu}$, were thought to be the only elementary particles.

In 1935 Yukawa predicted a new type of particle, called a meson, supposed to be responsible for the nuclear force. Anderson and Neddermeyer simultaneously produced experimental evidence for mesons in cosmic rays. Further work on these mesons in 1947 revealed, however, that they did not possess the properties which Yukawa's meson should possess. This first crisis in elementary particle physics was resolved in Powell's discovery in the same year, of two types of mesons in photographic emulsions; the Yukawa (π) meson and the lighter μ meson, now called the muon which was the one Anderson and Neddermeyer had found. They have very different properties.

As a starting point I shall assume that the reader has some familiarity with these particles which were the ones known in 1947: they will be discussed further in this chapter. Their properties appear in the table at the beginning of this book.[†]

1.2 UNITS AND KINEMATICS

For conservative reasons energy in elementary particle physics is measured in MeV. 1 MeV = 10^6 eV = 1.602×10^{-6} erg = 1.602×10^{-13} J. Since particles in cosmic rays and accelerators travel so fast the theory of relativity must be used to calculate momenta and masses, according to the well-known relation

---

[†] For a very readable account of the history of elementary particles, see for example C. N. Yang, *Elementary Particles*, Princeton University Press, 1961

ELEMENTARY PARTICLES AND SYMMETRIES

$$E^2 = p^2c^2 + m^2c^4 \tag{1-1}$$

where E is the energy, p the momentum and m the rest-mass of a particle, and c the velocity of light. Since c occurs so often it is usual to put $c = 1$ and quote all energies, momenta and masses in MeV. (Or, to be strict, energies in MeV, momenta in MeV/c and masses in MeV/c^2.) Thus the mass of the proton is

$$m_p = 938.21 \pm 0.01 \text{ MeV}/c^2$$

This means that $m_p c^2 = 938.21 \pm 0.01$ MeV. As accelerator energies have become higher, the energy unit

$$1 \text{ GeV} = 1 \text{ BeV} = 10^3 \text{ MeV}$$

has also come into use.

To see how relativistic kinematics work let us calculate the mass of π^+ from the decay $\pi^+ \to \mu^+ + \nu_\mu$, putting the suffix on ν_μ to emphasize that this neutrino is produced alongside a muon. We assume that

$$m_\mu = 105.65 \text{ MeV}/c^2, \quad m_{\nu_\mu} = 0$$

the first of which is accurately true and the second in more doubt.

According to equation (1) energy and momentum are not independent. If we write them as a four-vector using the convention of an imaginary fourth coordinate, $p = (\mathbf{p}, iE)$ (or (p_x, p_y, p_z, iE)), (1) gives (with $c = 1$)

$$\mathbf{p}^2 - E^2 = -m^2$$

or

$$p_x^2 + p_y^2 + p_z^2 - E^2 = -m^2 \tag{1-2}$$

We must choose a reference frame to work in. In this example let us choose the frame in which π^+ is at rest. Its energy-momentum is then

$$P_\pi = (0, 0, 0, im_\pi)$$

If the muon moves off along the positive z axis, the neutrino will move along the negative z axis, and

$$P_\mu = (0, 0, p_\mu, iE_\mu)$$

$P_\nu = (0, 0, p_\nu, iE_\nu)$

Conservation of energy-momentum gives

$$P_\pi = P_\mu + P_\nu$$

i.e.

$$m_\pi = E_\mu + E_\nu$$

$$p_\mu + p_\nu = 0; \quad p_\mu = -p_\nu = p, \tag{1-3}$$

and equation (2) gives

$$E_\mu = \sqrt{m_\mu^2 + p^2}, E_\nu = p, \tag{1-4}$$

when $m_\nu = 0$ has been used. Substituting (4) back into (3) gives

$$m_\mu = \sqrt{m_\mu^2 + p^2} + p$$

By deflecting μ^+ in a magnetic field, experiment gives

$$p = 29.80 \pm 0.04 \text{ MeV}/c,$$

so finally

$$m_{\pi^+} = \sqrt{(105.65)^2 + (29.80)^2} + 29.80$$

$$= 139.58 \pm 0.05 \text{ MeV}/c^2$$

It was assumed in this calculation that $m_\nu = 0$. In principle this assumption may be checked by comparing the calculated value of m_π with the value calculated from other experiments. Unfortunately, however, the value of m_π in this experiment is quite insensitive to m_{ν_μ}: even is $m_{\nu_\mu} = 2$ MeV, m_π changes by only 0.07 MeV, barely more than the experimental error. The experimental upper limit on the mass of ν_μ is 1.6 MeV.

EXERCISE Consider the reaction $\pi^- + p \to n + \gamma$. Assuming the π-meson is captured from an S-state Bohr orbit around p (and therefore has negligible kinetic energy) prove that the kinetic energy (=total energy − rest-mass energy) of the released neutron is 8.872 MeV.

1.3 CROSS SECTIONS AND INTERACTIONS

A great deal of information is gained by studying reaction cross sections between particles. These are quoted in units of area (1 barn, $b = 10^{-24}$ cm^2), and by the rules of quantum mechanics are related to the *probability* of the reaction occurring. Let us consider three typical processes.

a) $\pi^+ + p \rightarrow \pi^+ + p$

The cross section for this reaction is shown in Figure 1. The peak is about 200 mb, and σ tails off, for large pion kinetic energy, to tens of mb.

b) $\gamma + p \rightarrow \pi^+ + n$

This is called pion photoproduction. The cross section at the peak is ~ 200 μb. It is about 100–1000 times smaller than the cross section of reaction (a). It is typical of reactions involving γ.

c) $\nu_e + n \rightarrow e^- + p, \bar{\nu}_e + p \rightarrow n + e^+$

These reactions are shown in Figure 3. The cross sections are $\sim 10^{-14}$ b, about 10^{-12} times the cross section for (a). They are typical of reactions involving neutrinos. (The suffix on ν_e is to denote that the neutrino is involved in the reaction with e^- (or e^+). Particle and antiparticle conventions are explained below.)

The explanation given of these widely differing typical reaction cross sections is that there are *3 different types of interactions in nature*, called strong, electro-

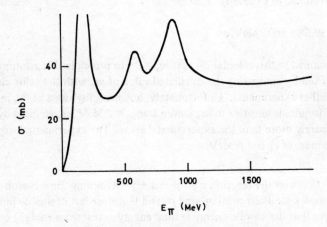

Figure 1.1 Total cross section for $\pi^+ + p$ scattering

INTRODUCTION

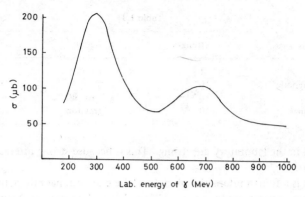

Figure 1.2   Cross section for $\gamma + p \to \pi^+ + n$

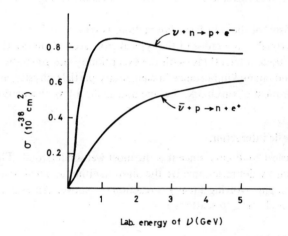

Figure 1.3 Theoretical cross sections for $\nu_e + n \to p + e^-$ and $\bar{\nu}_e + p \to n + e^+$. (From Lee and Yang, *Phys. Rev. Letters,* 4, 307, 1960)

magnetic and weak, and characterised by the cross sections in (a), (b) and (c) respectively. The electromagnetic interaction, of course, is not new. It was first recognised by Coulomb in the 18th century. The other two are 20th century discoveries. The following rules hold:

All reactions involving γ are electromagnetic
All reactions involving ν are weak
All reactions involving e are electromagnetic or weak
All reactions involving μ are electromagnetic or weak

All the other particles are, in general, capable of all 3 types or interaction — for example the proton is involved in (a), (b) and (c) above, but most reactions

Table 1.1

Interaction	Strength	Agent	Range
Strong	1	?	$\sim 10^{-13}$ cm
Electromagnetic	10^{-2}	γ	∞
Weak	10^{-12}	$-$ (or W)	0 (or $\hbar/m_W c$)
Gravitational	10^{-37}	graviton	∞

observed in the laboratory are strong. This is because strong interactions are so dominant.

There is a fourth interaction in nature, the gravitational interaction. It is the only one we apprehend in everyday life and was discovered by Newton in the 17th century. It is extremely weak — 10^{-25} times the strength of the weak interaction.

A comparison of the four basic interactions is set out in Table 1. The relative strengths are roughly the ratios of the typical cross sections in (a), (b) and (c). The study of these interactions (with the exception of the gravitational interaction) is of fundamental importance in elementary particle physics, and will occupy us for most of this book. Let me now make a few observations about each one.

Electromagnetic interaction

This is the easiest to discuss, since it is the most well understood. Electromagnetic interactions by definition involve the photon, either as a real or a virtual particle. An example involving a real γ is reaction (b) above. An example involving a virtual one is elastic e^-p scattering.

$$e^- + p \to e^- + p.$$

In classical physics we would say that these particles interact because the electron, say, "feels" the electromagnetic field of the proton, and this affects its motion. In quantum field theory, however, the electromagnetic field itself is quantised and becomes particle-like — a photon — and we then say that the electron and the proton "exchange" a photon. This is represented pictorially by the "Feynman diagram" of Figure 4. The photon is virtual because it is emitted and reabsorbed in such a short time that the uncertainty principle forbids its detection. Let the photon have energy E. Then it cannot be detected in a time interval less than $\tau \sim \hbar/E$, and it will then have travelled a distance $c\tau \sim c\hbar/E$. This is the range of the force. The higher the energy (and therefore momentum) that the photon carries, the smaller is the range of the force.

The minimum energy a particle may have is its rest-mass, which in the case of γ is zero. In this case, then, the range is infinite. It is this which makes electro-

INTRODUCTION

Figure 1.4 Electron-proton scattering mediated by photon exchange.

magnetic forces comparatively easy to discover. Two charged bodies exert a Coulomb force on each other no matter how far apart they are placed.

Other examples of reactions mediated by the electromagnetic interactions are

$$\pi^0 \to 2\gamma$$

$$\Sigma^0 \to \Lambda + \gamma$$

$$e^- + e^- \to e^- + e^-,$$

the last one of which is analogous to electron-proton scattering.

Strong interaction

Nuclear physicists call this the nuclear interaction, since it binds the nuclear particles together. It is obviously strong (since nuclei are so stable) and of short range (since nuclei of adjacent atoms do not "feel" each other). Yukawa's suggestion was that, in analogy with electromagnetism, there is also a particle exchanged between, say, p and n in a nucleus, to provide the strong force, as in Figure 5. If the mass of the exchanged particle (now called a pion π) is m_π, then the range of the force is $\hbar/m_\pi c$. If this is $\sim 10^{-13}$ cm, then $m_\pi \sim 180$ MeV.

As mentioned above, after some confusion the pion was found, and is indeed largely responsible for the strong force between protons and neutrons in nuclei, i.e. at low energy.

The pion, however, is not the universal agent of the strong interaction, as the photon is of the electromagnetic interaction. For example, it is impossible for the strong interaction between pions themselves to be mediated by exchange of

Figure 1.5 The strong interaction between proton and neutron mediated by pion exchange

pions: something else must be exchanged. Also, at high energies proton-proton and proton-neutron scattering are not well described by pion-exchange: again something else is needed.

The agent of strong interactions is, in fact, unknown at present. One possibility is that so-called "Regge poles" are exchanged. These objects are defined for all values (including unphysical ones) of the spin. When the spin becomes physical, the Regge pole corresponds to a real particle.[†]

Weak interaction

The classical model of weak interactions is due to Fermi, who suggested they are point interactions, of zero range. The beta-decay

$$n \to p + e^- + \bar{\nu}$$

is then described by the Feynman diagram of Figure 6. There are at present no experimental grounds for doubting Fermi's theory, but there are theoretical problems associated with it. These are less severe if the interaction is not a point interaction, but is carried by the exchange of an "intermediate boson" as in Figure 7, which shows how the boson W would mediate beta decay $n \to p + e^- + \bar{\nu}_e$. If the intermediate boson exists it can decay in a wide variety of ways, for example

$$W^\pm \to e^\pm + \nu_e$$
$$W^\pm \to \mu^\pm + \nu_\mu$$
$$W^\pm \to \pi^\pm + \pi^0.$$

Many experiments have been performed to look for the W in reactions such as

$$\nu_e + p \to p + e^- + W^+$$

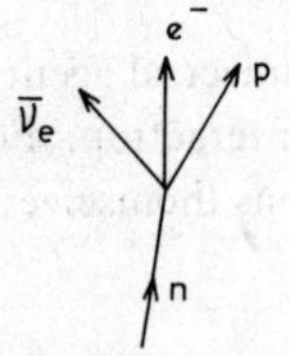

Figure 1.6 Fermi weak interaction for $n \to p + e^- + \bar{\nu}_e$.

[†] For more information on Regge poles, see for example D. H. Perkins, *Introduction to High Energy Physics*, Addison-Wesley, 1972, Chapter 7; S. Frautschi, *Regge Poles and S Matrix Theory*, W. A. Benjamin, 1963; or R. Omnès, *Introduction to Particle Physics*, Wiley-Interscience, 1971, Chapter 14.

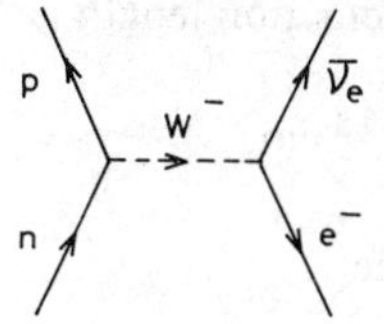

INTRODUCTION 9

Figure 1.7 Neutron decay $n \to p + e^- + \bar{\nu}_e$ mediated by an intermediate boson W

but a failure to find any indicates either that W does not exist or that it is too heavy to be produced in these reactions at present accelerator energies. Assuming W exists, the experiments can be used to put a lower limit on its mass of

$$m_W > 2 \text{ GeV} \tag{1-5}$$

Gravitational interaction

All particles, even the photon, gravitate. Assuming that nature is consistent, the gravitational field must also be quantised, and its quantum, the graviton, must be exchanged in gravitational interactions. The rest mass of this particle must be zero since the gravitational force has infinite range — and travels with the speed of light. A theory of quantised gravity is difficult to set up because the classical theory, due to Einstein, is non-linear (gravitational energy itself gravitates).

Weber claims recently to have observed gravitational waves coming from the centre of the galaxy.† If his claim is vindicated there is some remote but exciting hope that in the far future individual gravitons may be observed.

Because the gravitational force is so much weaker than the others it is neglected in studying elementary particles. We shall have no more occasion to mention it.

Physicists believe, of course, that eventually all four basic interactions will be integrated into a single comprehensive structure, but no-one supposes that this climactic advance is just round the corner.

1.4 INTERACTIONS AND DECAY TIMES

The three elementary particle interactions discussed above are not only responsible for two-body reactions, they also give rise to decays. In fact, the weak interaction was discovered through the beta decay $n \to p + e^- + \bar{\nu}_e$. Just as there are characteristic cross-sections for the reactions, there are also characteristic lifetimes for the decays. Taking the strong interactions first, if a typical cross section is

$$\sigma_{strong} \sim 10^{-25} \text{ cm}^2,$$

---

† See J. Weber, *Physical Review Letters* **22**, 1320 (1969); **24**, 276 (1970); **25**, 180 (1970).

then a corresponding strong interaction length is

$$l_{strong} = \sqrt{\sigma_{strong}} \sim 3 \times 10^{-13} \text{ cm},$$

and this is associated with a time

$$\tau_{strong} \sim \frac{l_{strong}}{c} \sim 10^{-23} \text{ sec}.$$

This is the order of magnitude of the time for strong interactions to take place. It is immensely short compared with, say, the lifetime of the Λ particle

$$\tau_\Lambda \sim 10^{-10} \text{ sec}$$

and we conclude that the Λ decays (into $p\pi^0$, $pe^-\bar{\nu}_e$, etc) must be weak. (The weaker the interaction, the longer time it takes, according to the probabilistic interpretation of quantum mechanics.) We already know that $\Lambda \to pe^-\bar{\nu}_e$ is weak, since $\bar{\nu}_e$ is involved: the purpose of the argument is to show that $\Lambda \to p\pi^-$ is also weak, *even though no neutrino is involved.*

10^{-10} sec is in fact a typical time for weak decays, and many such decays may be picked out in the table of particles. Times of this order are deduced directly by measuring the length of the particle track in a bubble chamber. If the particle is travelling, say, at $v = c/10$, then (since the time dilation is only very small for $v/c = 1/10$) the distance travelled in the bubble chamber before decay is

$$d = 3 \times 10^{10} \times 1/10 \times 10^{-10} = 0.3 \text{ cm}$$

and this track is of measurable length. For a decay time of 10^{-23} secs, however, a track measurement is impossible. On the other hand, if a particle decays in this very short time, it will have a measurable width (uncertainty in mass) given by

$$\Gamma \sim \frac{\hbar}{\tau} \sim \frac{6 \times 10^{-22} \text{ MeV sec}}{10^{-23} \text{ sec}} \sim 60 \text{ MeV}$$

This easily measurable: the first peak in Figure 1, for example, is due to the Δ particle whose width of 120 MeV is read straight from the graph. The widths of most of the higher mass particles ("resonances") are of this order of magnitude. They therefore decay by strong interactions.

Decays by electromagnetic interactions lie in the intermediate range; for example a small width compared with 10 MeV (e.g. 2.3 keV for η decay) or a short lifetime compared with 10^{-10} sec (e.g. 10^{-14} sec for Σ^0 decay).

These rules are only guidelines, for the decay time of a particle depends, as well as on the interaction strength, on the phase space available to the decay products. Neutron beta decay $n \to p + e^- + \bar{\nu}_e$ has a very long lifetime (1000 sec) compared with 10^{-10} sec. This is because there is so little energy to spare. The decay may be compared with parking a car in a crowded town on the Saturday afternoon before Christmas; it is possible in principle, but there is hardly any space available, so it takes a long time.

1.5 CLASSIFICATION OF PARTICLES

The particles $e^{\pm}, \nu_e, \bar{\nu}_e, \nu_\mu, \bar{\nu}_\mu$ and γ have no strong interactions. They all have spin ½ (in units of $\hbar$) except the photon, and are named as follows

Leptons $e^{\pm}, \mu^{\pm}, \nu_e, \bar{\nu}_e, \nu_\mu, \bar{\nu}_\mu$ spin ½

Photon γ spin 1

Particles with strong interactions are called *hadrons*

Hadrons $p, n, \pi, \Sigma, K, \ldots$ all spins $n/2$ (n integral)

All hadrons, of course, are either fermions or bosons and these are called respectively *baryons* and *mesons*

Baryons $p, n, \Sigma, \Lambda, \ldots$ spin ½

$\Delta, \Omega^-, \ldots$ spin 3/2

.
.
.
.
.
.

Mesons $\pi, K, \ldots$ spin 0

$\rho, \omega, K^*, \ldots$ spin 1

.
.
.
.
.

1.6 CONSERVATION OF CHARGE, BARYON NUMBER AND LEPTON NUMBER

The usefulness of the above classification depends on the conservation of class.

It has been implicitly assumed so far that particles may be labelled by their charge Q. It should be understood that this is because *charge is conserved in all reactions*. No reaction has ever been observed which changes the charge: for example $n \to p + \gamma$. If it were, the whole business of labelling would become ambiguous. There is an association

labelling of particles ↔ conservation of quantum numbers.

Now the proton is absolutely stable. In particular it does not decay into a positron

$$p \not\to e^+ + \gamma.$$

Why not? The answer must be that there is a law of *conservation of baryon number B* where

Baryons have $B = 1$

Mesons, leptons & γ have $B = 0$

There are no known reactions which violate this law: it is impossible to change the number of baryons in any reaction. A similar situation holds for leptons. For example,

$$e^- + e^- \not\to \pi^- + \pi^-.$$

This is evidence for *conservation of lepton number L* (where e^- has $L = 1$ and π^- has $L = 0$). But we may go further. The following reactions do *not* occur

$$\mu^\pm \not\to e^\pm + \gamma$$

$$\mu^\pm \not\to e^\pm + e^+ + e^-.$$

and a simple conservation of lepton number would not explain this. In addition

$$\pi^- \to \mu^- + \bar{\nu}_\mu, \pi^- \to e^- + \bar{\nu}_e,$$

but

INTRODUCTION

$\pi^- \not\to \mu^- + \nu_\mu, \pi^- \not\to e^- + \nu_e,$

and

$n \to p + e^- + \bar{\nu}_e$

but

$n \not\to p + e^- + \nu_e.$

(The neutrinos and antineutrinos differ in their helicity — see Chapter 8 — and this enables them to be distinguished experimentally). Also

ν_μ + nucleus $\to$ nucleus + μ^-

ν_μ + nucleus $\not\to$ nucleus + e^-

which was the experiment performed in 1962 showing that the neutrinos ν_e and ν_μ are distinct.

To explain all this we introduce two separate quantum numbers L_e and L_μ (electron number and muon number) with the following assignments

	L_e	L_μ
e^-	1	0
e^+	−1	0
ν_e	1	0
$\bar{\nu}_e$	−1	0
μ^-	0	1
μ^+	0	−1
ν_μ	0	1
$\bar{\nu}_\mu$	0	−1
other particles	0	0

L_e and L_μ are *both conserved* in all interactions. Of course $L = L_e + L_\mu$. Note that the muon decays according to

$\mu^- \to e^- + \bar{\nu}_e + \nu_\mu$

consistent with L_e and L_μ conservation.

EXERCISE 1)   Check that the assignments of L_e and L_μ explain all of the cited experimental facts.

14   ELEMENTARY PARTICLES AND SYMMETRIES

EXERCISE 2)   Which of the following decays do not occur? Why not? $\pi^- \to e^-e^+e^-$; $n \to p\mu^-\bar{\nu}_\mu$; $n \to \mu^+\mu^-\gamma$; $n \to \mu^+e^-\nu_e$; $\pi^+ \to \pi^0 e^+\nu_e$; $\pi^+ \to \pi^0 e^+\bar{\nu}_e$; $\pi^+ \to \mu^+\nu_\mu$; $\pi^0 \to e^+\nu_e$.

1.7 ANTIPARTICLES

The reader is doubtless familiar with the concept of antiparticles; to each particle there corresponds an antiparticle with the same mass and spin but with the values of Q, B, L_e and L_μ (and strangeness S — see Chapter 4) all *reversed*. In the table of leptons above both particles and antiparticles have been included. Particles with $Q = B = L_e = L_\mu = S = 0$ are their own antiparticles, for example γ and π^0. Particles and antiparticles annihilate each other to give states with $Q = B = L_e = S = 0$, for example

$$p + \bar{p} \to \pi^+ + \pi^- + 2\pi^0 \text{ (strong)}$$

$$e^+ + e^- \to 2\gamma \text{ (electromagnetic)}$$

FURTHER READING

Swartz, C. E. *The Fundamental Particles*, Addison-Wesley, 1965.
Perkins, D. H. *Introduction to High Energy Physics*, Addison-Wesley, 1972, Chapter 1.
Longo, M. J. *Fundamentals of Elementary Particle Physics*, McGraw-Hill 1973, Chapters 1 and 4.

CHAPTER 2

Symmetries and Conservation Laws

ONE OF THE main concerns in elementary particle physics over the last twenty years has been the study of symmetries. This has been partly because there has been, and still is, no satisfactory theory of strong and weak interactions, and a next best thing to knowing a theory is knowing what symmetry properties the theory has. Symmetries have also been studied because they have provided a surprisingly fruitful area of investigation: isospin, unitary symmetry and Ω^-, quarks, parity violation, CP violation and current algebra are all to do with symmetries. As a background to discussing these topics in later chapters, it is appropriate to make some general observations on symmetry here.

2.1 SYMMETRIES IN CLASSICAL PHYSICS

Symmetry in physics is a consequence of *unmeasurability* or *indistinguishability*. As an example, there is no absolute origin in space: we may choose the origin of the coordinate system to be wherever we like, and the laws of physics will be the same. In other words, absolute position is unmeasurable. What consequence does this have?

Consider an isolated system of two particles interacting only with each other. If their coordinates are r_1 and r_2, the potential energy of interaction is

$$V(r_1, r_2),$$

a function of the positions of the two particles. Now, if there is no absolute origin in space, we may measure the coordinates from a different origin, obtained from the first one by translation through a distance $-a$, say.† Measured in this system the potential energy is

$$V(r_1 + a, r_2 + a).$$

---

† Or, alternatively, we may move to a new position by translation through a.

16 ELEMENTARY PARTICLES AND SYMMETRIES

But if the two origins are equivalent, the energies must be the same:

$$V(\mathbf{r}_1, \mathbf{r}_2) = V(\mathbf{r}_1 + \mathbf{a}, \mathbf{r}_2 + \mathbf{a}) \text{ for all } \mathbf{a}. \tag{2-1}$$

A function of two variables which obeys an equation such as this must be a function of the *difference* $\mathbf{r}_1 - \mathbf{r}_2$ only. For example, we may have $V(\mathbf{r}_1, \mathbf{r}_2) = (\mathbf{r}_1 - \mathbf{r}_2)^2$ or $V(\mathbf{r}_1, \mathbf{r}_2) = \exp i\mathbf{k}\cdot(\mathbf{r}_1 - \mathbf{r}_2)$. Both of these obey (1). If, on the other hand, $V(\mathbf{r}_1, \mathbf{r}_2) = (\mathbf{r}_1 + \mathbf{r}_2)^2$, then $V(\mathbf{r}_1 + \mathbf{a}, \mathbf{r}_2 + \mathbf{a}) = (\mathbf{r}_1 + \mathbf{r}_2 + 2\mathbf{a})^2$ and (1) does not hold. So we have

$$V(\mathbf{r}_1, \mathbf{r}_2) = V(\mathbf{r}_1 - \mathbf{r}_2). \tag{2-2}$$

Now force **F** is given by

$$\mathbf{F} = -\frac{\partial V}{\partial \mathbf{r}}$$

so the forces on the two particles are

$$\mathbf{F}_1 = -\frac{\partial V}{\partial \mathbf{r}_1}, \quad \mathbf{F}_2 = -\frac{\partial V}{\partial \mathbf{r}_2}$$

and the total force acting on the system is

$$\mathbf{F} = \mathbf{F}_1 + \mathbf{F}_2 = -\frac{\partial V}{\partial \mathbf{r}_1} - \frac{\partial V}{\partial \mathbf{r}_2}. \tag{2-3}$$

However, because of (2),

$$\frac{\partial V}{\partial \mathbf{r}_1} = \frac{\partial V}{\partial (\mathbf{r}_1 - \mathbf{r}_2)}, \quad \frac{\partial V}{\partial \mathbf{r}_2} = -\frac{\partial V}{\partial (\mathbf{r}_1 - \mathbf{r}_2)}$$

therefore

$$\frac{\partial V}{\partial \mathbf{r}_1} = -\frac{\partial V}{\partial \mathbf{r}_2},$$

and from (3)

$$\mathbf{F} = 0; \tag{2-4}$$

SYMMETRIES AND CONSERVATION LAWS

the total force acting on two particles which interact only with each other is zero. What does this imply? According to Newton's second law of motion

$$\mathbf{F} = \frac{d\mathbf{p}}{dt} \qquad (2\text{-}5)$$

where **p** is momentum. Equation (4) and (5) now give

$$\frac{d\mathbf{p}}{dt} = 0; \quad \mathbf{p} = \text{constant}. \qquad (2\text{-}6)$$

In other words, the momentum of the system is *conserved*. We may generalise this example to any number of particles and say that the momentum of an isolated system of particles is conserved.

By means of this example, we have been able to see the connection between (a) *unmeasurability* of absolute position, (b) *invariance* under spatial translation (equation (1)), and (c) *conservation* of momentum (equation (6)).

The connection between these concepts is very general: it applies in quantum as well as in classical physics, and it applies to other unmeasurable quantities than the origin of space. Let us first consider the general formulation in classical physics. A system is described by a Lagrangian L which is a function $L(q_j, \dot{q}_j, t)$ of the generalised positions q_j of the particles, labelled by j; their time derivatives $\dot{q}$, and time t. Lagrange's equations are[†]

$$\frac{d}{dt}\left(\frac{\partial L}{\partial \dot{q}_j}\right) - \frac{\partial L}{\partial q_j} = 0.$$

If the coordinate q is not absolutely measurable, then L does not depend on q_j: $\partial L/\partial q_j = 0$, so Lagrange's equations give

$$\frac{d}{dt}\left(\frac{\partial L}{\partial \dot{q}_j}\right) = 0; \quad \frac{\partial L}{\partial \dot{q}_j} = \text{constant}.$$

However $\partial L/\partial \dot{q}_j = p_j$, the generalised momentum, so we have p_j = constant; unmeasurability of generalised position implies conservation of generalised momentum. If q is position, p is momentum; and if q is an angle, p is angular momentum. So as well as the result obtained above, that invariance under spatial displacements leads to conservation of linear momentum, we see that invariance

---

† See, e.g. H. Goldstein, *Classical Mechanics*, Addison-Wesley, 1950, Chapter 1.

under *rotations* (which follows from the *unmeasurability of absolute direction* in space) leads to *conservation of angular momentum*. (Equivalence of all directions in space is not at all obvious, even to the unpreoccupied mind. It would seem that "sideways" is not equivalent to "up" or "down". Newton, however, realised that this was due to gravity, not to an intrinsic property of space. This was the significance, for Newton, of the falling apple.) Finally, since there is no absolute origin in time, all systems must be invariant under time displacements. L may not depend on t:

$$\frac{\partial L}{\partial t} = 0.$$

However, since $\partial L/\partial t = -\partial H/\partial t$[†], where H is the Hamiltonian, this gives $\partial H/\partial t = 0$ and

H = constant.

H, however, is the total energy, so we obtain conservation of energy. These invariances and conservation laws are summarised in Table 1. They are applicable (as far as anyone knows) to all physical laws, and therefore must be symmetries of strong, weak and electromagnetic interactions.

2.2 SYMMETRIES IN QUANTUM PHYSICS

Let us now turn to the quantum theory. Here a system is described by a wave function $\psi(x, t)$ which obeys the Schrödinger equation[‡]

$$H\psi = i\hbar \frac{\partial}{\partial t} \psi \tag{2-7}$$

where H is the Hamiltonian, or energy operator. The complex conjugate equation is

$$H\psi^* = -i\hbar \frac{\partial}{\partial t} \psi^*. \tag{2-8}$$

Consider an observable Q. Its expectation value at time t is

[†] See Goldstein, op.cit., Chapter 7.
[‡] See, e.g., R. M. Eisberg, *Fundamentals of Modern Physics*, Wiley, 1961, p. 321.

SYMMETRIES AND CONSERVATION LAWS

Table 2.1.
Some Symmetries (Exact and Inexact) of Nature

Unmeasurability of	implies invariance under	and conservation of	extent to which symmetry holds
Absolute position in space	space translation	momentum	exact
Absolute time	time displacement	energy	exact
Absolute direction in space	rotations	angular momentum	exact
Relative phase of charged and neutral particles	charge gauge transformations	charge Q	exact
Relative phase of baryons and other particles	baryon gauge transformations	baryon number B	exact
Relative phase of e^- & ν_e and other particles	electron number gauge transformation	electron number L_e	exact
Relative phase of μ^- & ν_μ and other particles	muon number gauge transformations	muon number L	exact
(Indistinguishability of) left and right	space inversion P	parity	violated in weak interactions
Direction of time flow	time reversal T	–	violated but source unknown
Distinction between particles and antiparticles	charge conjugation	charge parity	violated in weak and perhaps also elmag. interactions

$$\overline{Q}(t) = \int \psi^*(t) Q \psi(t) d^3x.$$

The rate of change of $\overline{Q}$ is, from (7) and (8)

$$i\hbar \frac{\partial}{\partial t} \overline{Q}(t) = \int \psi^*(t)(QH - HQ)\psi(t) d^3x$$

$$= \int \psi^*(t)[Q, H]\psi(t)d^3x \tag{2-9}$$

where

$$[Q, H] = QH - HQ \tag{2-10}$$

is called the *commutator* of Q and H. If the expectation value of Q is conserved, then

$$\bar{Q} \text{ conserved: } \frac{\partial \bar{Q}}{\partial t} = 0 \Rightarrow [Q, H] = 0, \tag{2-11}$$

and Q and H are said to *commute*.

Consider now an infinitesimal displacement of the coordinate system along the x axis through a distance $-\delta x$. The wave function becomes

$$\psi(x) \to \psi(x + \delta x) = \left(1 + \delta x \frac{\partial}{\partial x}\right)\psi(x)$$

$$= \left(1 + \frac{i}{\hbar}\delta x p_x\right)\psi(x) \tag{2-12}$$

where we have expanded by a Taylor series to first order in δx, and where

$$p_x = \frac{\hbar}{i}\frac{\partial}{\partial x} \tag{2-13}$$

is called the *generator* of translations along the x axis. It will be recognised as being the momentum operator in quantum theory.[†] We now want to show that invariance under infinitesimal space translations (along the x axis) implies conservation of (the x component of) momentum. To do this, note that if the system is invariant under $x \to x + \delta x$, then the expectation value of H is unchanged

$$\int \psi^*(x) H \psi(x) d^3x = \int \psi^*(x + \delta x) H \psi(x + \delta x) d^3x$$

$$= \int \psi^*(x)\left(1 - \frac{i}{\hbar}\delta x p_x\right) H \left(1 + \frac{i}{\hbar}\delta x p_x\right)\psi(x)d^3x.$$

---

† See, e.g., Eisberg, op.cit., p.203.

SYMMETRIES AND CONSERVATION LAWS

This implies that

$$H = \left(1 - \frac{i}{\hbar}\delta x p_x\right) H \left(1 + \frac{i}{\hbar}\delta x p_x\right)$$

$$= H - \frac{i}{\hbar}\delta x [p_x, H] + 0(\delta x)^2$$

i.e.

$$[p_x, H] = 0; \quad \overline{p_x} \text{ conserved} \tag{2-14}$$

where (11) has been used. This finishes the proof. By performing the same computation for $y \to y + \delta y$, $z \to z + \delta z$, we find that invariance under space translations implies conservation of momentum, just as in classical theory. Similarly, invariance under time displacements gives conservation of energy.

Lastly consider angular momentum. An infinitesimal rotation about the z axis through $\delta\varphi$ gives

$$x' = x + y\delta\varphi \equiv x + \Delta x$$

$$y' = y - x\delta\varphi \equiv y + \Delta y$$

$$z' = z \equiv z + \Delta z. \tag{2-15}$$

The wave function becomes

$$\psi(\mathbf{r}) \to \psi(\mathbf{r}) + \Delta\mathbf{r} \cdot \nabla\psi(\mathbf{r}) + \ldots$$

$$= \psi + \Delta x \frac{\partial\psi}{\partial x} + \Delta y \frac{\partial\psi}{\partial y} + \Delta z \frac{\partial\psi}{\partial z} + \ldots$$

$$= \left[1 + \left(y\frac{\partial}{\delta x} - x\frac{\partial}{\delta y}\right)\delta\varphi\right]\psi + \ldots$$

$$= \left[1 - \frac{i}{\hbar}L_z\delta\varphi\right]\psi + \ldots \tag{2-16}$$

where

$$L_z = \frac{\hbar}{i}\left(x\frac{\partial}{\partial y} - y\frac{\partial}{\partial x}\right) \tag{2-17}$$

22   ELEMENTARY PARTICLES AND SYMMETRIES

is the *generator* of rotations about the z axis. It will be recognised as being the operator for the z component of angular momentum in quantum thoery.[†] The rest of the proof goes through as for momentum. That is, invariance under rotations about the z axis implies conservation of L_z;

$$[L_z, H] = 0, \overline{L_z} \text{ conserved.} \tag{2-18}$$

And, of course, invariance under rotations about all axes implies conservation of angular momentum;

$$[\mathbf{L}, H] = 0; \overline{\mathbf{L}} \text{ conserved.} \tag{2-19}$$

It is convenient to write down the finite forms of (12) and (16). They are

translation along x axis: $\quad \psi \to \exp(ixp_x/\hbar)\psi$

arbitary translation: $\quad \psi \to \exp(i\mathbf{x}\cdot\mathbf{p}/\hbar)\psi \tag{2-20}$

rotation about z axis: $\quad \psi \to \exp(-i\varphi L_z/\hbar)\psi$

arbitary rotation: $\quad \psi \to \exp(-i\boldsymbol{\varphi}\cdot\mathbf{L}/\hbar)\psi \tag{2-21}$

From the existence of three unmeasurable quantities we have derived, in classical and in quantum theory, the existence of three conserved quantities. It is worth mentioning that they are of a very fundamental kind. They express the facts that the laws of physics must be the same in all places, at all times and be independent of orientation in space. If these invariances did *not* hold then the laws of nature would vary in time and from place to place. Experiments would not be repeatable, and science itself may be impossible.

2.3 GAUGE TRANSFORMATIONS AND OTHER SYMMETRIES

We have already discussed, in Chapter 1, four conservation laws not connected with space and time: electric charge Q, baryon number B, electron number L_e and muon number L_μ. These follow from invariance of physical laws under so-called gauge transformations; these express the fact that the phase of the wave function of (in the case of charge) a charged particle relative to a neutral one, is unmeasurable. Invariance under gauge transformations leads to a conserved current, and thereby to a conserved charge.

The conservation laws for these four charges have in common with conservati

---

[†] See, e.g., Eisberg, op.cit., p.314.

SYMMETRIES AND CONSERVATION LAWS

of momentum, energy and angular momentum the fact that they are all *absolute*. They are *exact* symmetries in the sense that they are symmetries of all the interactions; strong, electromagnetic and weak. This is in striking contrast with the symmetries to be discussed later. These are of two types. First, there are symmetries related to space inversion and time reversal, and second, there are symmetries completely unrelated to space and time; they include isospin, strangeness and unitary symmetry. All of these symmetries are not observed by at least one interaction; they are called *approximate symmetries*, and will occupy us for most of the rest of this book.

2.4 SPIN OF THE CHARGED PION

Since angular momentum is conserved, all elementary quantum systems must be in eigenstates of angular momentum. The intrinsic angular momentum of elementary particles is called their spin. Let us discuss the experimental determination of the spin of the charged pion. It is found by comparing the rates for the reactions $\pi^+ + d \to p + p$ and its inverse $p + p \to \pi^+ + d$, where d is the deuteron.

Assuming that a reaction $a + b \to c + d$ is invariant under time reversal (see sections 2.11 and 2.12), a principle called the *principle of detailed balance* holds, according to which

$$\frac{\frac{d\sigma}{d\Omega}(a+b \to c+d)}{\frac{d\sigma}{d\Omega}(c+d \to a+b)} = \left(\frac{p_f}{p_i}\right)^2 \frac{(2s_c+1)(2s_d+1)}{(2s_a+1)(2s_b+1)}$$

where s_i is the spin of i, and p_i and p_f are the initial and final relative momenta in the centre of mass system in the reaction $a + b \to c + d$. Since p and d have $s = \frac{1}{2}$ and $s = 1$ respectively, we have

$$\frac{\frac{d\sigma}{d\Omega}(\pi^+ + d \to p+p)}{\frac{d\sigma}{d\Omega}(p+p \to \pi^+ + d)} = \frac{p^2}{q^2} \cdot \frac{4}{3(2s_{\pi^+}+1)}$$

where p and q are the centre of mass momenta of the proton and pion respectively. The cross sections for both processes have been measured at various momenta and give the result $s_{\pi^+} = 0$. Since π^- is the antiparticle of π^+ it also has spin zero.

2.5 SPIN OF THE NEUTRAL PION

π^0 is also believed to have spin zero, though the evidence is more indirect. It comes from the decay

$$\pi^0 \to 2\gamma.$$

Firstly, it can be shown that if π^0 has $s = 1$, this decay is forbidden. This leaves $s = 0, s \geqslant 2$. The rest of the analysis is concerned with finding the parity as well as the spin. We shall be considering parity later, but here I need only mention that the parity of bosons is either $+1$ or -1.

The reasoning now goes as follows. The two photons in the decay will be polarised — each one either left- or right-circularly. Also, each photon decays into an electron-positron pair

$$\pi^0 \to \gamma \underset{\hookrightarrow e^+ e^-}{} + \gamma \underset{\hookrightarrow e^+ e^-}{} ,$$

each pair production defining a plane. It can be shown that if π^0 has spin-parity 0^+, these planes are parallel, whereas if it has 0^- they are perpendicular. Experimentally they are found to be perpendicular, indicating spin-parity 0^-. Spin $\geqslant 2$ is not rigorously excluded by this argument, but there is not serious enough doubt to make it worth pursuing the matter further. Besides, the success of isospin (see Chapter 3) demands that π^0 have the same spin as $\pi^\pm$, i.e. zero.†

2.6 SPACE INVERSION AND PARITY

General remarks

Although parity is one of the key concepts in elementary particle physics, it does not rest on the same foundation that spin does. The existence of spin follows from the conservation of angular momentum, and this is guaranteed by invariance under rotations. Space inversion, however, defined by

$$\begin{pmatrix} x \\ y \\ z \\ t \end{pmatrix} \xrightarrow{P} \begin{pmatrix} -x \\ -y \\ -z \\ t \end{pmatrix} \qquad (2\text{-}22)$$

---

† For further details on the $\pi^0 \to 2\gamma$ decay, see for example I. S. Hughes, *Elementary Particles*, Penguin, 1972, p. 63; or G. Källén, *Elementary Particle Physics*, Addison—Wesley, 1964, p.50.

SYMMETRIES AND CONSERVATION LAWS

where P is the parity operator, is not a rotation. This may easily be seen by observing that a rotation through π about the z axis, say, results in (see Figure 1)

$$\begin{pmatrix} x \\ y \\ z \\ t \end{pmatrix} \xrightarrow{R_z(\pi)} \begin{pmatrix} -x \\ -y \\ z \\ t \end{pmatrix} \tag{2-23}$$

which differs from (22) in that only two coordinates are reversed. Space inversion invariance, then, does not follow from invariance under rotations.† Nature, in fact, taking advantage of this loophole, is not completely invariant under parity, since weak interactions are not, as we shall see in Chapter 8. Neglecting the weak interactions, then, consider the Schrödinger equation

$$H(\mathbf{x})\psi(\mathbf{x}, t) = i \frac{\partial \psi(\mathbf{x}, t)}{\partial t} \tag{2-24}$$

where H is the Hamiltonian. If H is invariant under $\mathbf{x} \to -\mathbf{x}$,

$$H(\mathbf{x}) = H(-\mathbf{x}). \tag{2-25}$$

Substitute $\mathbf{x} \to -\mathbf{x}$ in (24)

$$H(-\mathbf{x})\psi(-\mathbf{x}, t) = i \frac{\partial \psi(-\mathbf{x}, t)}{\partial t}.$$

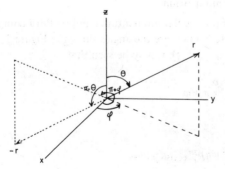

Figure 2.1

---

† A. Pais, in a lecture in Copenhagen in July 1963, recorded that he asked Dirac why he had not introduced parity in his famous book on quantum mechanics, "and Dirac with characteristic simplicity answered: 'Because I did not believe in it.'"

Using (25) this gives

$$H(\mathbf{x})\psi(-\mathbf{x}, t) = i\frac{\partial \psi(-\mathbf{x}, t)}{\partial t},$$

so $\psi(-\mathbf{x}, t)$ obeys the same equation as $\psi(\mathbf{x}, t)$. We may put

$$\psi(-\mathbf{x}, t) = \psi'(\mathbf{x}, t) = P\psi(\mathbf{x}, t) \tag{2-26}$$

where P is the parity operator; and the combinations

$$\psi_0(\mathbf{x}, t) = \frac{1}{\sqrt{2}}(\psi(\mathbf{x}, t) - \psi'(\mathbf{x}, t))$$

$$\psi_E(\mathbf{x}, t) = \frac{1}{\sqrt{2}}(\psi(\mathbf{x}, t) + \psi'(\mathbf{x}, t)) \tag{2-27}$$

have odd and even parity

$$P\psi_0 = -\psi_0, \quad P\psi_E = \psi_E.$$

The reader may also verify that $[H, P] = 0$, so that parity is conserved. Parity is the quantum number whose conservation results from invariance under space inversion.

Parity and angular momentum

The wave function for angular momentum l and its third component m is given by $Y_l^m(\theta, \varphi)$[†], where θ and φ are the angles shown in Figure 1. Under space reflection $(x, y, z) \to (-x, -y, -z)$, it may be seen that

$$\theta \xrightarrow{P} \pi - \theta, \quad \varphi \xrightarrow{P} \pi + \varphi.$$

Now

$$Y_l^m(\theta, \varphi) \propto \sin^{|m|}\theta \, P_l^m(\cos\theta) e^{im\varphi}$$

where $P_l^m(\cos\theta)$ is the associated Legendre polynomial of order $(\cos\theta)^{l-|m|}$. So, under space reflection

---

[†] See, for example, Eisberg, op.cit., pp.301, 304.

SYMMETRIES AND CONSERVATION LAWS

$$Y_l^m(\theta,\varphi) \to Y_l^m(\pi-\theta,\pi+\varphi) = (+1)^{|m|}(-1)^{l-|m|}(-1)^m Y_l^m(\theta,\varphi)$$

$$= (-1)^l Y_l^m(\theta,\varphi),$$

so the parity of a state of orbital angular momentum l is $(-1)^l$.

Intrinsic parity

A state consisting of particles a and b will be described by the wave function

$$\Psi_{ab} = \psi_a \psi_b.$$

If the intrinsic parities of a and b are ϵ_a and ϵ_b, then

$$P\psi_a = \epsilon_a \psi_a, \quad P\psi_b = \epsilon_b \psi_b,$$

and if the orbital angular momentum between a and b is l, then

$$P\Psi_{ab} = \epsilon_a \epsilon_b (-1)^l \Psi_{ab}.$$

Now imagine that particle c is *created* by scattering a and b

$$a + b \to a + b + c$$

and let the total angular momentum of the right hand side be L. Then, if this reaction conserves parity

$$\Psi_{ab}^* P\Psi_{ab} = \Psi_{abc}^* P\Psi_{abc}$$

i.e.

$$\epsilon_a \epsilon_b (-1)^l = \epsilon_a \epsilon_b \epsilon_c (-1)^L$$

i.e.

$$\epsilon_c = (-1)^{L+l}$$

which gives the intrinsic parity of c. If, however, it is not possible to create c alone, its intrinsic parity cannot be defined. If, for instance, c is a fermion, or if it has electric charge, it must be created alongside another particle d, and the above reasoning will give a value for $\epsilon_c \epsilon_d$ — it will be impossible to find ϵ_c alone. So *intrinsic parity is only well defined for particles which can be created singly i.e. for uncharged (and nonstrange) bosons.*

Relative parity

On a moment's reflection it is seen that the limited usefulness of intrinsic parity arises because of *conservation laws*. Charge conservation forbids creation of a single charged particle by itself, and therefore determination of its intrinsic parity. If there are two charged particles, c^+ and d^+, however, such that it is possible to have the reaction

$$a + c^+ \to a + d^+,$$

then if the initial and final orbital angular momenta are l_1 and l_2, conservation of parity gives

$$\epsilon_a \epsilon_c (-1)^{l_1} = \epsilon_a \epsilon_d (-1)^{l_2}$$

i.e.

$$\frac{\epsilon_c}{\epsilon_d} = (-1)^{l_1 + l_2},$$

and the *relative parity* of c^+ and d^+ may be found. Exactly the same holds for baryon number as for charge. (It also holds for strangeness, since the interactions which change the strangeness, and would therefore allow comparison of the parities of strange and non-strange particles, are the weak interactions, which violate parity!, and are therefore useless.)

It follows from these considerations that the parity of all baryons may be defined only relative to the parity of (say) p, n, and Λ. A convenient way of representing this is to adopt the

convention: p, n and Λ have positive parity. (2-28)

The examples below will illustrate how this works out in practice.

Particles and Antiparticles

It can be shown that for

$$\begin{cases} \text{bosons:} & \text{particle and antiparticle have the same parity} \\ \text{fermions:} & \text{particle and antiparticle have opposite parity} \end{cases} \quad (2\text{-}29)$$

Transformations under P

From the definition $\mathbf{r} \xrightarrow{P} -\mathbf{r}$, $t \xrightarrow{P} t$, it follows that momentum $\mathbf{p} \xrightarrow{P} -\mathbf{p}$, since $\mathbf{p} = m\dot{\mathbf{r}}$. Angular momentum (and therefore also spin) does not change sign, since $\mathbf{l} = \mathbf{r} \times \mathbf{p}$. These and other transformation properties to be discussed later are listed in Table 2.

SYMMETRIES AND CONSERVATION LAWS

Table 2.2

Quantity	P transform	T transform	C transform
r	−r	r	r
t	t	−t	t
p (momentum)	−p	−p	p
l = r × p (angular momentum)	l	−l	l
σ (spin)	σ	−σ	σ
E	−E	E	−E
B	B	−B	−B
Q	Q	Q	−Q

2.7 PARITY CONSERVATION IN STRONG AND ELECTROMAGNETIC INTERACTIONS

The classic test for parity conservation in strong interactions is that of Tanner, which concerns the reaction

$$p + F^{19} \to O^{16} + \alpha$$

in the energy region corresponding to an excited state of Ne^{20} known to have spin-parity = 1^+. If, in this energy region, the reaction is actually

$$p + F^{19} \to Ne^{20*} \to O^{16} + \alpha$$
$$\phantom{p + F^{19} \to Ne^{20*}} 1^+ 0^+ 0^+$$

then the resonance will show up in the spectrum of α. Since O^{16} and α both have spin$^P = 0^+$, the final state has $0^+, 1^-, 2^+, \ldots$, corresponding to $l = 0, 1, 2, \ldots$. Conservation of J then demands 1^-, and since Ne^{20*} has 1^+, the reaction is forbidden if P is conserved. In this case no resonance will show up in the α-spectrum. The upper limit for parity violation in this strong interaction was quoted by Tanner to be

$$|F|^2 < 4 \times 10^{-8},$$

where F is the relative strength of parity violating and parity conserving amplitudes. More recent similar experiments bring this figure down to $|F|^2 < 10^{-12}$.

If there are parity violating interactions among hadrons, there will be a corresponding *parity impurity* in the states: that is, states will not have a simple parity, but will contain a slight admixture of states of the opposite parity. The upper limit on F quoted above already excludes such an admixture due to electromagnetic interactions, and this proves that electromagnetic interactions con-

30 ELEMENTARY PARTICLES AND SYMMETRIES

serve parity. In fact the limit of 10^{-12} is already in the realm of weak interactions (see Table 1.1) which *do* violate parity. It is therefore impossible to improve on the limit of 10^{-12}.

2.8 PARITY OF THE PHOTON

The photon is described by the 4-vector potential $(\mathbf{A}, i\varphi) = A_\mu$, where $\mathbf{A}$ is the vector, and φ the scalar potential. The magnetic and electric fields $\mathbf{B}$ and $\mathbf{E}$ are given by

$$\mathbf{B} = \text{curl}\mathbf{A}, \quad \mathbf{E} = -\text{grad}\varphi - \frac{\partial}{\partial t}\mathbf{A} \tag{2-30}$$

The electric field may be produced by a pair of oppositely charged plates in the xy plane at $z = z_0/2$ and $z = -z_0/2$. Under parity $z \to -z$, and the plates change places, thus changing the direction of $\mathbf{E}$:

$$\mathbf{E} \xrightarrow{P} -\mathbf{E}.$$

Since the operator $\text{grad} \equiv \mathbf{i}(\partial/\partial x) + \mathbf{j}(\partial/\partial y) + \mathbf{k}(\partial/\partial z)$ changes sign under parity, this implies from (30) that A_μ behaves like

$$(\mathbf{A}, i\varphi) \xrightarrow{P} (-\mathbf{A}, i\varphi) \tag{2-31}$$

i.e. A_μ behaves like an ordinary vector, rather than a pseudovector; the space part of A changes sign and the *photon has negative intrinsic parity*.

EXERCISE   Using (31) prove that $\mathbf{B}$ behaves like a pseudovector under parity. Check that this corresponds to the physical picture of $\mathbf{B}$ as being generated by circulating charges.

2.9 PARITY OF THE CHARGED PION

The parity of π^- may be determined by using the fact that the reaction

$$\pi^- + d \to 2n \tag{2-32}$$

takes place, in which π^- is captured from an S-orbit around the deuteron d. The parity of the left hand side is

$$\epsilon(lhs) = \epsilon(\pi^-)\epsilon(d)(-1)^l = \epsilon(\pi^-)\epsilon(d),$$

SYMMETRIES AND CONSERVATION LAWS

since $l = 0$. d is a bound state of p and n with orbital angular momentum zero (with a small admixture of $l = 2$), so using the convention (28), $\epsilon(d) = +1$ and

$$\epsilon(lhs) = \epsilon(\pi^-). \tag{2-33}$$

On the other hand, d has spin 1, so $(\pi^- d)$ has $J = 1$, and by conservation of angular momentum the $2n$ state also has $J = 1$. Two neutrons obey the Pauli exclusion principle, and the allowed states are (cf. Table 3-3)

$S = 1$ (symmetric) and l odd (antisymmetric)
$S = 0$ (antisymmetric) and l even (symmetric).

$S = 0$ and l even, however, gives J even, and in our case $J = 1$. The two neutrons therefore have $S = 1$, $l = 1$ $(J = 1)$ and

$$\epsilon(2n) = [\epsilon(n)]^2(-1)^1 = -1$$

which compared with (33), assuming parity is conserved, gives $\epsilon(\pi^-) = -1$. Since the pion is a boson, π^+ has the same parity, so, subject to the convention (28), the charged pion has negative parity.

2.10 PARITY OF THE NEUTRAL PION

In contrast to the reaction (32), the reaction

$$\pi^- + d \to 2n + \pi^0 \tag{2-34}$$

seems to be highly forbidden. Using the above analysis it is easy to see that if π^0 and π^- have the same parity (-1), reaction (34) will take place, but with π^0 being emitted in an $l = 1$ state relative to the $2n$. If π^- and π^0 have opposite parities, however, it will take place with $l = 0$ and therefore will be much faster. The fact that the observed rate is very slow means that π^0 has negative (intrinsic) parity.

It is also of interest that the reaction $\pi^0 \to 2\gamma$, by which the π^0 spin was found to be 0, also gave its parity as -1. This may, alternatively, be viewed as confirmation that this *electromagnetic interaction* conserves parity.

In conclusion, pions have spin-parity $J^P = 0^-$. They are *pseudoscalar* particles — see Table 3.

2.11 TIME REVERSAL

In analogy with equation (22), time reversal is defined by

Table 2.3.

Field/particle	Behaviour under P	Type	J^P	Example
spin 0	$\varphi \to \varphi$	scalar	0^+	$\pi_N(1016)$
spin 0	$\varphi \to -\varphi$	pseudoscalar	0^-	π
spin 1	$(\mathbf{V}, V_4) \to (-\mathbf{V}, V_4)$	vector	1^-	γ
spin 1	$(\mathbf{V}, V_4) \to (\mathbf{V}, -V_4)$	pseudovector/ axial vector	1^+	$B(1220)$

$$\begin{pmatrix} x \\ y \\ z \\ t \end{pmatrix} \xrightarrow{T} \begin{pmatrix} x \\ y \\ z \\ -t \end{pmatrix} \qquad (2\text{-}35)$$

A quantum system characterised by its Hamiltonian H obeys the Schrödinger equation

$$H\psi(\mathbf{x}, t) = i \frac{\partial \psi(\mathbf{x}, t)}{\partial t}. \qquad (2\text{-}36)$$

Even if H is invariant under time reversal, $H \xrightarrow{T} H$, however, the corresponding equation in the time reversed system is

$$H\psi(\mathbf{x}, -t) = -i \frac{\partial \psi(\mathbf{x}, -t)}{\partial t}. \qquad (2\text{-}37)$$

which is *not* the same as (36). $\psi(\mathbf{x}, t)$ and $\psi(\mathbf{x}, -t)$ do not obey the same equation. The quantity which obeys the same equation as $\psi(\mathbf{x}, -t)$ is $\psi^*(\mathbf{x}, t)$, for the complex conjugate of (36) is

$$H^*\psi^*(\mathbf{x}, t) = -i \frac{\partial \psi^*(\mathbf{x}, t)}{\partial t}$$

and we can therefore put $\psi(\mathbf{x}, -t) = \psi^{*\prime}(\mathbf{x}, t)$. The fact that T changes a wave function into its complex conjugate means that a state cannot be in an eigenstate of T, and therefore that *there is no corresponding conserved quantum number*. This is a fundamental difference between space inversion and time reversal.

SYMMETRIES AND CONSERVATION LAWS

Until 1964 all physical laws were thought to be invariant under time reversal, that is, the arrow of time *on a microscopic scale* was thought to be unmeasurable. (On a *macroscopic* scale, there *is* an arrow of time, since systems always evolve towards a more chaotic state.) The observation of the decay $K_2^0 \to 2\pi$, however, provided indirect evidence that T invariance is not a universal law of nature. This will be discussed in Chapter 9.

2.12 TIME REVERSAL INVARIANCE IN STRONG AND ELECTRO-MAGNETIC INTERACTIONS

Since time reversal relates a wave function $\psi_a(t)$ to its complex conjugate $\psi_a^*(-t)$, it relates the transition matrix element $\psi_a^* M \psi_\beta$ to $\psi_\beta^* M \psi_a$. In other words, it relates the amplitude for $a \to \beta$ to that for $\beta \to a$. A way of testing T invariance is to compare the rates for processes which can be performed either "forwards" or "backwards", for example

$$^{24}Mg + d \leftrightarrow {}^{25}Mg + p$$

$$^{24}Mg + a \leftrightarrow {}^{27}Al + p. \tag{2-38}$$

Note that this type of test, called the principle of *detailed balance,* does not specify equal transition rates for the forward and backward reactions. The transition rate W from a to β is given by Fermi's Golden Rule[†]

$$W_{a \to \beta} = \frac{2\pi}{\hbar} |M_{a\beta}|^2 \rho_\beta$$

where $M_{a\beta}$ is the matrix element and ρ_β is the density of states in β. According to the principle of detailed balance, $|M_{a\beta}| = |M_{\beta a}|$, but $W_{a \to \beta}$ and $W_{\beta \to a}$ will differ as the phase space factors ρ_β and ρ_a differ.

The reactions (38) have been used to show that T invariance holds to better than 0.5%. This, of course, is a test for strong interactions.

Tests of detailed balancing in electromagnetic interactions have also been carried out on the reactions

$$n + p \leftrightarrow d + \gamma. \tag{2-39}$$

The results show agreement with detailed balancing, but only to an accuracy of about 20%. The most spectacular test for T-invariance in electromagnetic inter-

---

[†] See, for example, Eisberg, op.cit., p.291.

actions is the non-existence of a *neutron dipole moment*. The neutron, although neutral, may have a positive and negative charge distribution associated with it. If the "centres of gravity" of the positive and negative charges are not at the same point but separated by a distance **d**, this results in a dipole moment *f*e**d**, where *f* is a number specifying the magnitude of the effect. In what direction would **d** lie? The only other intrinsic vector associated with a neutron is its spin σ, so we would expect **d** and σ to be parallel. However, the presence of the quantity e**d**.σ is evidence of *T* violation, for (see Table 2)

$$f e\mathbf{d}.\sigma \xrightarrow{T} -fe\mathbf{d}.\sigma \qquad (2\text{-}40)$$

T invariance requires

Tinv: $fe\mathbf{d}.\sigma = -fe\mathbf{d}.\sigma$, $f = 0$.

Very accurate experiments have been performed in recent years by Ramsey and collaborators to look for a neutron dipole moment. The present upper limit is

$$\mu_{el.\,dip.}(n) = (2 \pm 2) \times 10^{-23} e.cm. \qquad (2\text{-}41)$$

There is a slight difficulty of interpretation with this test, since, as well as (40), we have

$$f e\mathbf{d}.\sigma \xrightarrow{P} -fe\mathbf{d}.\sigma$$

so *the existence of a dipole moment is also evidence that parity is violated*. Electromagnetic interactions, however, conserve parity, so it may be that $f = 0$ is guaranteed independently of *T* invariance. The weak interactions, however, violate parity, so a dipole moment may arise from an interplay between the interactions, the weak interactions violating *P* and the electromagnetic ones *T*. In that case, *f* must be a product

$$f = (G_F M_p^2)g$$

where G_F is the Fermi coupling constant (see section 10.2), $G_F M_p^2 \approx 10^{-5}$ is a dimensionless number typifying the weak interactions, and *g* is a parameter specifying the degree to which electromagnetic interactions violate *T*. The magnitude of the dipole moment would then be

$$\mu_{el.\,dip.}(n) \sim f.\lambda_p e$$

where $\lambda_p = \hbar/M_p c$ is the Compton wavelength of the proton = 2×10^{-14} cm.

SYMMETRIES AND CONSERVATION LAWS

This gives

$$\mu_{el.dip.}(n) \sim g.10^{-5} \times 2 \times 10^{-14} e.cm.$$

$$= 2g \times 10^{-19} e.cm.$$

Comparison with (41) gives $g < 10^{-4}$. This is the accuracy to which T invariance holds in electromagnetic interactions, as measured by this test.

2.13 CHARGE CONJUGATION

Besides P and T, which are related to space and time, we may define an operation C which replaces every particle by its antiparticle

C: particle $\leftrightarrow$ antiparticle.

C therefore changes the sign of all charges (electric charge, baryon and lepton numbers and strangeness), and so $\mathbf{E} \xrightarrow{C} -\mathbf{E}$, $\mathbf{B} \xrightarrow{C} -\mathbf{B}$, since $\mathbf{E}$ and $\mathbf{B}$ are due to static and moving charges. On the other hand, all spacetime quantities are unchanged (see Table 2).

Eigenstates of C must have zero charge, for example π^0, $p\bar{p}$, γ, e^+e^-. For these states, we may define a "charge parity" ϵ_c

$$C\varphi = \epsilon_c \varphi,$$

and since C^2 is clearly the identity,

$$\epsilon_c^2 = 1, \quad \epsilon_c = \pm 1.$$

Since $\mathbf{E} \xrightarrow{C} -\mathbf{E}$, $\mathbf{B} \xrightarrow{C} -\mathbf{B}$, then

$$A_\mu = (\mathbf{A}, i\varphi) \xrightarrow{C} -A_\mu,$$

and the photon has negative charge parity

$$C\gamma = -\gamma: \quad \epsilon_c(\gamma) = -1. \tag{2-42}$$

As will be seen below, to quite a good accuracy strong and electromagnetic interactions conserve C, so it makes sense to determine the charge parity of π^0 from the decay $\pi^0 \to 2\gamma$. In fact, it follows directly from (42) that

$$C\pi^0 = \pi^0: \quad \epsilon_c(\pi^0) = +1. \tag{2-43}$$

2.14 C CONSERVATION IN STRONG AND ELECTROMAGNETIC INTERACTIONS

Many predictions of C invariance are difficult to test, since they relate a reaction involving particles to one involving antiparticles, and it is not always easy to get hold of antiparticles. For instance, it is not easy to compare $\pi^- p$ scattering with $\pi^+ p$ scattering, because of a natural deficiency of antiprotons. The best way to test for C invariance is to use states which are C-eigenstates, for example $p\bar{p}$. In fact, in proton-antiproton annihilation

$$p + \bar{p} \to \pi^+ + \pi^- + \pi^0 + \ldots,$$

if C invariance holds, the angular distribution of π^+ and π^- must be the same. This prediction has been checked, and C invariance found to hold to better than 1%.

As for electromagnetic interactions, T. D. Lee and coworkers have in the last few years thrown doubt on C conservation (see Sections 9.8 and 9.9). So far, however, there is no convincing evidence that C is violated. A convenient way to test for C violation is to look for an asymmetry in the π^+ and π^- distributions in

$$\eta \to \pi^+ \pi^- \pi^0 \tag{2-44}$$

which, from its lifetime, is seen to be an electromagnetic decay. A recent (1968) experiment gives the asymmetry as $1.5 \pm 0.5\%$. We cautiously conclude that electromagnetic interactions are invariant under C, to about this level of accuracy.

2.15 A NOTE ON LAWS OF NATURE AND BOUNDARY CONDITIONS

It is well known, at least to chemists, that there are many "optically active" organic compounds, which rotate the plane of polarisation of light to the right or left. For example, the glucose which occurs in nature is dextro-rotatory. It is important to realise that this is *not* evidence for parity violation, for laevo-rotator glucose can be made in the laboratory and is otherwise identical to the dextro-rotatory variety. The reason that only one of them is found in nature is an illustration of the part played by chance in the early stages of evolution of the environment on earth. The continued dominance of one variety throughout evolution is due to the fact that two right handed molecules react in a way in which a right and a left handed molecule cannot react, just because of the geometry of the molecules. Thus the asymmetry is due to a lack of symmetry in the initial conditions rather than to any intrinsic asymmetry in the laws of nature.

A similar remark may be made about C invariance. The fact that the earth (and probably the solar system) is made of protons, neutrons and electrons, rather than antiprotons, antineutrons and positrons, is not evidence for C violation. *C* invariance means that particles and antiparticles react in the same way, not that there are the same number of them in any place. The fact that the earth is made of p, n and e^- is a consequence of the initial conditions obtaining at the time of formation of the planet.

The description of any physical system involves a theory, or law of nature, and a knowledge of the boundary conditions characterising the particular problem. In the examples above, the boundary conditions were simply the initial conditions. The interesting thing is that the boundary conditions are outside the realm of any theory; there can be no theory to explain them. It is very strange, as Wigner has remarked, that our knowledge of the physical world divides so clearly into these two groups.†

2.16 SYMMETRIES AND GROUPS

It is, as the reader will appreciate, axiomatic that symmetry transformations must have the property that two successive transformations of a particular type are equivalent to one transformation of the same type. If for some bizarre reason two space translations

$$x' = x + a; \quad x'' = x' + b \tag{2-45}$$

were *not* equivalent to one overall translation

$$x'' = x + (a + b), \tag{2-46}$$

then we could not speak of invariance under space translations.

Transformations which have this simple property are said to form a *group*, which is defined as follows. A group G is a set of elements (physical operations in our case)

$$G = \left\{ A, B, C, \ldots \right\}$$

with respect to which a single law of composition (or multiplication) is defined. The product AB of the elements A and B must satisfy the following conditions (the symbol ϵ means "belongs to"):

1) Closure: If $A \epsilon G$, $B \epsilon G$, then $AB = C \epsilon G$.

2) Associative law: If $A \epsilon G$, $B \epsilon G$, $C \epsilon G$, then

$$(AB)C = A(BC)$$

---
† See E. P. Wigner, Nobel prize acceptance speech, in *The Nobel Lectures*, Elsevier, 1964. Reprinted in E. P. Wigner, *Symmetries and Reflections*, MIT Press, 1972.

where A, B and C are arbitary. Either side may then be denoted ABC.

3) Unit element: There exists an element $I \epsilon G$, called the identity or unit element, such that for every $A \epsilon G$

$$AI = IA = A$$

4) Inverse elements: For every element $A \epsilon G$, there exists an element $A^{-1} \epsilon G$ such that

$$AA^{-1} = A^{-1}A = I.$$

Mathematically, the consequences of a symmetry are determined by the structure of the group. One primary distinction that can be made is whether a group is *abelian* (or *commutative*) or not. A group of transformations is abelian if two successive transformations give the same result *no matter in what order they are performed*. An example is the group of translations, mentioned above. Instead of first translating through a, and then through b, as in (45) above, perform these the other way round. We then have

$$x_1 = x + b; \quad x_2 = x_1 + a$$

i.e.

$$x_2 = x + (a + b) = x'';$$

the final result does not depend on the order. The group of space translations (and for the same reason, of time displacements) is abelian. Similarly, a translation along the x axis followed by one along the y axis clearly gives the same result if they are performed the other way round; the reader is left to verify this. So the group of all translations in three dimensional space is *abelian* or *commutative*.

The translations are represented in quantum theory by the factors $\exp(ip_x/\hbar)$, $\exp(ip_y/\hbar)$ and $\exp(ip_z/\hbar)$, as in equation (13). p_x, p_y and p_z are said to be the *generators* of space translations. Since translations commute, the generators also commute. Define the *commutator* of two generators p_x and p_y by

$$[p_x, p_y] = p_x p_y - p_y p_x. \qquad (2\text{-}47)$$

Then the statement that the generators commute is simply

$$[p_x, p_y] = [p_y, p_z] = [p_z, p_x] = 0 \qquad (2\text{-}48)$$

SYMMETRIES AND CONSERVATION LAWS

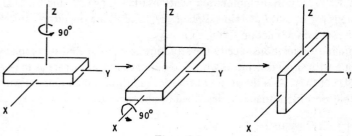

Figure 2.2

Rotations are slightly less simple. They do form a group, because two arbitary rotations can be shown to be equivalent to one rotation about a suitably chosen axis and through a suitably chosen angle. Rotations about different axes, however, *do not commute*. It is easy to demonstrate this.[†] Perform the following rotations on a book. (1) Rotate first 90° about the z axis and then 90° about the x axis (Figure 2). (2) Rotate first 90° about the x axis and then 90° about the z axis (Figure 3). The results are completely different.

Since rotations are represented in quantum theory by equation (19) and similar equations involving L_x and L_y for rotations about the x and y axes, then L_x, L_y and L_z, the *generators of rotations*, do not commute. It can be shown[‡] that they obey the relations (with $\hbar = 1$)

$[L_x, L_y] = iL_z$

$[L_y, L_z] = iL_x$

$[L_z, L_x] = iL_y$. (2-49)

These are called *commutation relations*.

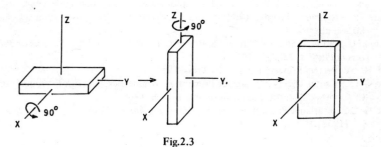

Fig.2.3

---

† The following demonstration is taken from R. P. Feynman, *Theory of Fundamental Processes*, Benjamin, 1962, Chapter 3.

‡ See, e.g., Feynman, op.cit., Chapter 3.

EXERCISE   Verify the commutation relations (49) using (20) and the analogous expressions for L_x and L_y. (Hint: since L_x, L_y and L_z are operators, proceed by operating (49) on a function $f(x, y, z)$.)

Commutation relations specify the structure of a group almost completely. The group with three generators whose commutation relations are as in (49) is called SU_2 or SO_3. It is the group of rotations in three dimensions. It will be found that isospin invariance is described by the same group.

EXERCISE   Prove that

$$[p_x, L_y] = ip_z \text{ and cyclic} \tag{2-50}$$

These commutation relations express the fact that (p_x, p_y, p_z) behave as the three components of a vector under rotations.

As a final example, Lorentz boost transformations are an example of transformations which do not form a group. Two successive boosts along different directions are not equivalent to a single boost, but involve rotations as well. The full Lorentz group, then, contains both boosts and rotations. Its structure is quite complicated. It will be seen in chapter 12 that according to the quark model of chiral symmetry, chiral transformations also have the property that by themselves they do not form a group. Isospin transformations must be included as well.

FURTHER READING

Feynman, R. P., R. B. Leighton and M. Sands, *The Feynman Lectures on Physics*, vol. III, Addison-Wesley, 1965, Chapter 17. (Symmetries and conservation laws in quantum theory.)

Gürsey, F., "Introduction to Group Theory" in *Relativity, Groups and Topology* (C. DeWitt and B. DeWitt, Eds.) Blackie, London; Gordon and Breach, New York, 1964.

Skinner, R., *Mechanics*, Blaisdell 1969, section 3.4. (Symmetry and conservation laws in classical physics.)

Rowe, E. G., and E. J. Squires, "Present Status of C, P and T invariance", *Reports on Progress in Physics*, 32, 273 (1969).

Weyl, H., *Symmetry*, Princeton University Press, 1952.

Wick, G. C., "Group Theory. Invariance Principles. Symmetries." in *High Energy Physics* (C. Dewitt & M. Jacobs, eds.), Gordon & Breach, 1965.

Wick, G. C., "Invariance Principles of Nuclear Physics", *Annual Reviews of Nuclear Science*, 8, 1 (1958).

CHAPTER 3

Isospin

3.1 INTRODUCTION: A HYPOTHESIS

AT THE BRIEFEST glance at a table of elementary particles, we can see straight away that the neutron and proton are very similar. Their masses are m_p = 938.3 MeV and m_n = 939.6 MeV; they both have spin ½ and baryon number B = 1, but their electric charges (and magnetic moments) are different. We know that they both experience the strong (nuclear) force and the electromagnetic force; and we also know that the strong force is a lot stronger than the electromagnetic one. So we make a *hypothesis: if they had the same charge (and magnetic moment) they would have the same mass.* In other words, their difference in mass is *due* solely to their different charge, and consequently to their different electromagnetic interactions. In saying this we have to assume that it has some meaning to say "if they had the same charge". They don't. But if the electromagnetic interactions are *independent* of the strong interactions, then we may imagine a fictitious world in which these interactions may be "turned on" and "turned off" independently of one another. If we then turn off the electromagnetic interactions, and thereby neutralise all the attractions and repulsions due to charges and magnetic moments, then, we are saying, the proton and neutron would be indistinguishable.

To look at it from another point of view, we may say that in creating the physical world, we turn on the strong interactions, and the proton and neutron appear with their nuclear interaction, and they are degenerate. We then turn on the electromagnetic interaction, the proton develops an electric charge and corresponding magnetic moment, the neutron doesn't; and their masses split apart (with n being heavier than p). In the light of this, we may rephrase the hypothesis:

As far as the strong interactions are concerned, the proton and neutron are identical. In the actual world, they are distinguishable only by their differing electromagnetic interactions.†   (3-1)

---

† It should be clear that it is possible to imagine this short excursion from the real world to a fictitious one only because the electromagnetic interactions are a lot weaker than the strong ones, and so may be considered as a small perturbation. In other words, it is possible to imagine what the world would look like without electromagnetic interactions, but without the strong interactions it would look unimaginably different.

3.2 EVIDENCE FOR THE HYPOTHESIS

The ideal way of verifying the hypothesis (1) is to calculate the changes in mass of p and n when the charges and magnetic moments are taken into account, and to compare this with the actual mass difference. Unfortunately, no-one has yet successfully performed this calculation. We must therefore look elsewhere for evidence.

As a first rough indication that the hypothesis is true, consider the quantity

$$\frac{\text{mass difference}}{\text{mean mass}},$$

which we would, on heuristic grounds, expect to be roughly equal to the relative strengths of the electromagnetic and strong interactions (tacitly assuming, that is, that the strong interactions are responsible for the bulk of the mass of elementary particles). The relative strengths are $\sim 10^{-2} - 10^{-3}$ (See Chapter 1), and

$$\frac{m_n - m_p}{\frac{1}{2}(m_n + m_p)} = \frac{1.3}{939} \approx 10^{-3},$$

so the hypothesis is borne out, to within an order of magnitude.[†]

[†] The odd thing about the neutron and the proton is that the neutron, the uncharged member, is the heavier. This would seem to tell against electromagnetism as being the only agent responsible for the breaking of isospin symmetry, since electromagnetism would surely make the *charged* particle heavier. However, as pointed out by Feynman and Speisman (*Physical Review* **94** 500 (1954)), the situation is not that simple, since the neutron and proton have *anomalous magnetic moments* as well as charge, and will interact with the electromagnetic field through these magnetic moments. Since the anomalous magnetic moment of the neutron ($-1.91\mu_N$) is numerically greater than that of the proton ($1.79\mu_N$), it is indeed quite conceivable that the overall effect of electromagnetism is to give the neutron a greater mass (greater self-energy) than the proton. Unfortunately, Feynman and Speisman were unable to calculate what the mass difference should be, since according to the standard methods of calculation, the self-energies concerned turn out, like all self-energies, to be infinite! In order to get a finite answer, one has arbitrarily to "cut off" the integral, but this means introducing an extra free parameter viz. the top limit of integration.

Widening our horizons somewhat, and taking the point of view that *all* hadrons, not just the proton and neutron, come in charge multiplets (see Chapter 4), we see from inspection of the table at the beginning of the book that as far as the stable *baryons* are concerned (i.e. $n, p; \Sigma^-, \Sigma^0, \Sigma^+; \Xi^0, \Xi^-$) there is the rule that the more negatively charged the particle, the heavier it is. This may be significant. However, for mesons there is not even a simple rule of thumb: the charged pion is heavier than the neutral one, but with the kaons it is the other way round! The whole subject of electromagnetic mass differences is a very vexing one, in urgent need of solution.

ISOSPIN

The most direct piece of evidence, however, comes from the study of strong interactions in nuclei. The nuclear force has two properties which we now discuss, and then consider their relation to the hypothesis (1).

3.3 CHARGE SYMMETRY OF NUCLEAR FORCES

The first property concerns so-called "mirror nuclei", that is, pairs of nuclei in which the numbers of protons and neutrons are interchanged, and so the nuclei may be converted into one another by $n \leftrightarrow p$. For example, H^3 and He^3, shown in Figure 1, are mirror nuclei.

Assuming the existence of only two-body forces, in both nuclei there are three pairs of interactions. In H^3 they are *p-n, p-n* and *n-n*, and in He^3 they are *p-n, p-n* and *p-p*. The difference, then, is that H^3 has a *n-n* force where He^3 has a *p-p* force. If we ignore the interactions due to magnetic moments, the *n-n* force is a purely strong force, but the *p-p* force has a strong part and an electrostatic part (Coulomb part). We may write

$$V_{nn} = (V_{st})_{nn}$$

$$V_{pp} = (V_{st})_{pp} + (V_{Coul})_{pp}$$

in an obvious notation, where V stands for potential.

We now study the properties of mirror nuclei, noting in advance that *any differences between two mirror nuclei will be due to $(V_{Coul})_{pp}$ and the difference between $(V_{st})_{nn}$ and $(V_{st})_{pp}$*. The binding energies of H^3 and He^3 are

H^3: B.E. = 8.492 MeV

He^3: B.E. = 7.728 MeV.

We may calculate the Coulomb energy in He^3 due to $(V_{Coul})_{pp}$:—

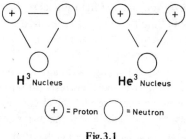

Fig.3.1

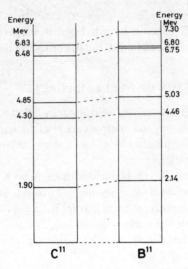

Figure 3.2   Energy levels of the nuclei C^{11} and B^{11}

$$\tfrac{1}{2}Z(Z-1) \times \tfrac{6}{5} \frac{e^2}{4\pi\epsilon_0 R} \quad (\text{where } R = 1.45 \times 10^{-15}(A)^{1/3} m.)$$

$$= \frac{6 \times (1.6 \times 10^{-19})^2}{5 \times 1.45 \times (3)^{1/3} \times 10^{-15} \times 4\pi \times 8.85 \times 10^{-12}} J.$$

$= 1.32 \times 10^{-13} J.$

$= 0.826$ MeV.

The actual difference in binding energy is 0.764 MeV. This is quite close to 0.826 MeV, especially when it is remembered that the quoted formula gives a value for R which is slightly too small in light nuclei such as He^3. We may conclude that the difference in binding energy is accounted for *solely by* $(V_{Coul})_{pp}$.

A more sensitive test is to look at the energy levels of a pair of mirror nuclei. For example, consider the low-lying levels of B^{11} and C^{11} shown in Figure 2. Here the ground states are assigned energy zero, and it is seen that the energies of the various excited levels in the two nuclei correspond fairly closely. The differences in energy may again be explained by the charge on the proton.

From evidence such as this, we deduce that the difference between two mirror nuclei is accounted for by $(V_{Coul})_{pp}$ only, and hence that

ISOSPIN

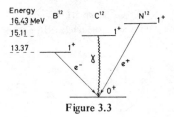

Figure 3.3

$(V_{st})_{nn} = (V_{st})_{pp}.$

This is called the *charge symmetry of nuclear forces*.

3.4 CHARGE INDEPENDENCE OF NUCLEAR FORCES

Consider the two (mirror) nuclei B^{12} and N^{12}. They each have ground states with spin-parity 1^+, which decay to the ground state of C^{12}; B^{12} by emission of an electron, N^{12} by emission of a positron. The spin-parity of the C^{12} ground state is 0^+. The energy difference between the ground states of B^{12} and C^{12} is 13.37 MeV, and that between the ground states of N^{12} and C^{12} 16.43 MeV. This state of affairs is shown in Figure 3. As expected from the charge symmetry of nuclear forces, there is a close similarity between the ground states of B^{12} and N^{12} which have respectively $5p$ and $7n$, and $7p$ and $5n$. By contrast, the ground state of C^{12}, with $6p$ and $6n$ has a very different energy, and different quantum numbers. However, on examination we find that C^{12} has an excited state at 15.11 MeV with spin-parity 1^+, which normally decays to the ground state by photon emission. This level, therefore, has an energy between the ground state energies of B^{12} and C^{12}. What does this mean?

Let us look at the number of forces between pairs of particles in these nuclei. In B^{12}, with $5p$ and $7n$, there are $5!/(3!2!) = 10$ *p-p* forces, $7!/(5!2!) = 21$ *n-n* forces and $7 \times 5 = 35$ *p-n* forces. Similar figures are obtained for C^{12} and N^{12}, and are set out in Table 1. Also in Table 1 are noted the values of $Z(Z-1)$ which is proportional to the Coulomb energy of the *p-p* repulsion. Comparing the three nuclei, what we see is that as the Coulomb energy increases from B^{12} to N^{12}, the energy of the level increases roughly similarly, so we conclude that the remaining forces are all equal, i.e.

$(V_{st})_{pp} = (V_{st})_{nn} = (V_{st})_{pn}$ in the same state. (3-2)

The qualification "in the same state" must be added because both the $^1S_0$ and $^3S_1$ states are accessible to *n-p* whereas only the $^1S_0$ state is accessible to *p-p* and *n-n* because of the Pauli exclusion principle. (This is of course why the ground state of C^{12} is a state which has no parallel in B^{12} or N^{12}.) The interaction energies may only be compared when the particles are in the same state.

ELEMENTARY PARTICLES AND SYMMETRIES

Table 3.1

p-p, *n-n* and *p-n* forces in the nuclei B^{12}, C^{12} and N^{12}. Also noted is the value for the nuclei

	pp	nn	pn	total	Z(Z − 1)
B^{12}	10	21	35	66	20
C^{12}	15	15	36	66	30
N^{12}	21	10	35	66	42

The condition (2), which clearly includes, but is stronger than charge symmetry, is called the *charge independence of nuclear forces*. Since the number of *p-n* interactions in C^{12} is only marginally more than the number in N^{12} and B^{12}, this does not make very compelling evidence for charge independence. However, in many other triads of nuclei a parallel situation has been found to exist. That is, if we take two mirror nuclei with $n \pm 1$ neutrons and $n \mp 1$ protons, we find similar energy levels (which may or may not be the ground state) — this is charge symmetry. If we now look at the nucleus in between, with n neutrons and n protons, we find the corresponding energy level in an intermediate position. Examples of triads for which this is true are He^6, Li^6; Be^6, C^{14}, N^{14}, O^{14}; O^{18}, F^{18}, Ne^{18}.

Besides this evidence from the static properties of nuclei, there is other evidence for charge independence from low energy scattering data. For details, I refer the reader to a recent review,[†] and merely state that from the scattering data, charge independence is shown to hold to within about 0.8%.

We conclude that the interactions between protons and neutrons are indeed charge independent, and so may consider the hypothesis (1) as justified — in a world with strong interactions only, the proton and neutron would be identical.

3.5 THE NUCLEON AND ISOSPIN

As originally observed by Heisenberg, since the proton and neutron are so similar we may describe them as two states of a single particle, the *nucleon N*. In analogy with spin, we introduce the quantity *isotopic spin* (nowadays commonly abbreviated to *isospin* or *I*-spin, and denoted *I*), and say that the nucleon has isospin $I = ½$. Its two component states with I_3, the component of *I* along the third axis, given by $I_3 = ½$ and $I_3 = -½$, are *p* and *n* respectively. This has a direct parallel with the fact that the electron has spin ½, so its spin projection along the z axis is +½ or −½.

[†] E. M. Henley, in *Isospin in Nuclear Physics*, ed. D. H. Wilkinson. North-Holland, 1969.

ISOSPIN

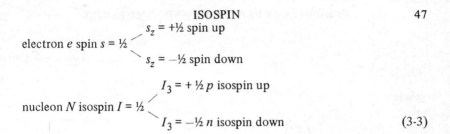

$$\text{electron } e \text{ spin } s = \tfrac{1}{2} \begin{cases} s_z = +\tfrac{1}{2} \text{ spin up} \\ s_z = -\tfrac{1}{2} \text{ spin down} \end{cases}$$

$$\text{nucleon } N \text{ isospin } I = \tfrac{1}{2} \begin{cases} I_3 = +\tfrac{1}{2}\, p \text{ isospin up} \\ I_3 = -\tfrac{1}{2}\, n \text{ isospin down} \end{cases} \tag{3-3}$$

There is clearly a connection between I_3 and electric charge Q for the nucleon

$$N: Q = I_3 + \tfrac{1}{2}. \tag{3-4}$$

It is important to realise that there is no new physics in the above; it is purely a formalism. The physics will come later. For the present, since the analogy between isospin and ordinary spin is so seductive, and will be often referred to, it is important to bear in mind the differences between them. One difference, the most obvious one, is that spin, being angular momentum, is related to behaviour under rotations in ordinary three dimensional space. Isospin, however, is related to behaviour in a purely abstract space, whose third axis according to equation (4) is related to charge. (It is for this reason that we call it the third axis, rather than the z axis.) There are other differences which we shall come across later.

Wigner pointed out that isospin is a useless idea unless the dominant forces are charge independent; unless, that is, the two components of the nucleon (with different charge) interact in the same way. We have seen above that this *is* true of their strong interactions, though not of their electromagnetic ones. What is the analogue of this situation in the case of spin? Consider the electron in the hydrogen atom. When there is no magnetic field the spin up and spin down states are degenerate. When a magnetic field is applied, however, the states acquire different energies and the degeneracy is broken. (This is what leads to the Zeeman effect.) The atom in zero magnetic field is equivalent, in the nucleon case, to a world in which there are strong interactions only (degenerate case). The atom in a non-zero magnetic field is equivalent to a world with strong and electromagnetic interactions (non-degenerate case). The *limitation* of the analogy is that in the case of the atom, the magnetic field may be turned on or off at will. The electromagnetic interactions, however, which remove the p/n degeneracy, are, like the poor, always with us. The world in which they are "turned off" is purely unreal.

3.6 ISOSPIN TRANSFORMATIONS

The wave function for the nucleon may be expressed as a two-component function, as in the case of spin ½.

$$\psi_N = \begin{pmatrix} a \\ b \end{pmatrix} = ap + bn \tag{3-5}$$

where

$$p = \begin{pmatrix} 1 \\ 0 \end{pmatrix}, \quad n = \begin{pmatrix} 0 \\ 1 \end{pmatrix}.$$

If we introduce the matrices

$$\tau_1 = \begin{pmatrix} 0 & 1 \\ 1 & 0 \end{pmatrix}, \quad \tau_2 = \begin{pmatrix} 0 & -i \\ i & 0 \end{pmatrix}, \quad \tau_3 = \begin{pmatrix} 1 & 0 \\ 0 & -1 \end{pmatrix} \tag{3-6}$$

(which are the Pauli spin matrices), then we see that

$$\tau_3 p = \begin{pmatrix} 1 & 0 \\ 0 & -1 \end{pmatrix} \begin{pmatrix} 1 \\ 0 \end{pmatrix} = \begin{pmatrix} 1 \\ 0 \end{pmatrix} = p$$

$$\tau_3 n = \begin{pmatrix} 1 & 0 \\ 0 & -1 \end{pmatrix} \begin{pmatrix} 0 \\ 1 \end{pmatrix} = \begin{pmatrix} 0 \\ -1 \end{pmatrix} = -n.$$

From (4) we require $I_3 p = \frac{1}{2} p$, $I_3 n = -\frac{1}{2} n$, so we identify

$$I_3 = \tfrac{1}{2} \tau_3 \tag{3-7}$$

for the nucleon. Similarly we define

$$I_1 = \tfrac{1}{2} \tau_1, \quad I_2 = \tfrac{1}{2} \tau_2. \tag{3-8}$$

Now the operators

$$I_+ \equiv I_1 + iI_2 = \tfrac{1}{2}(\tau_1 + i\tau_2) = \begin{pmatrix} 0 & 1 \\ 0 & 0 \end{pmatrix} \tag{3-9}$$

$$I_- \equiv I_1 - iI_2 = \tfrac{1}{2}(\tau_1 - i\tau_2) = \begin{pmatrix} 0 & 0 \\ 1 & 0 \end{pmatrix} \tag{3-10}$$

have the following effect on p and n states:

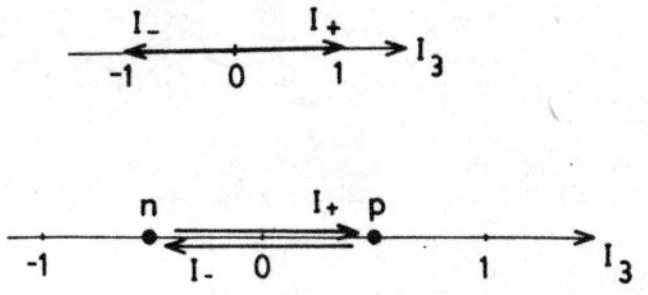

ISOSPIN

Figure 3.4 The raising and lowering operators I_+ and I_- and their effect on the nucleon states, p and n

$$I_+ p = \begin{pmatrix} 0 & 1 \\ 0 & 0 \end{pmatrix} \begin{pmatrix} 1 \\ 0 \end{pmatrix} = 0 \tag{3-11}$$

$$I_+ n = \begin{pmatrix} 0 & 1 \\ 0 & 0 \end{pmatrix} \begin{pmatrix} 0 \\ 1 \end{pmatrix} = \begin{pmatrix} 1 \\ 0 \end{pmatrix} = p \tag{3-12}$$

$$I_- p = \begin{pmatrix} 0 & 0 \\ 1 & 0 \end{pmatrix} \begin{pmatrix} 1 \\ 0 \end{pmatrix} = \begin{pmatrix} 0 \\ 1 \end{pmatrix} = n \tag{3-13}$$

$$I_- n = \begin{pmatrix} 0 & 0 \\ 1 & 0 \end{pmatrix} \begin{pmatrix} 0 \\ 1 \end{pmatrix} = 0. \tag{3-14}$$

I_+ changes n (with $I_3 = -\frac{1}{2}$) into p (with $I_3 = +\frac{1}{2}$) and I_- changes p into n: they are called for this reason *raising and lowering operators* respectively. They are shown diagramatically in Figure 4. Their matrix representations for the nucleon are τ_+ and τ_-; in fact τ_1, τ_2 and τ_3 are the matrix representation of I_1, I_2 and I_3 for $I = \frac{1}{2}$ states.

Now consider a rotation about the I_2 axis through an angle θ, as in Figure 5. The isospin wave function $\binom{a}{b}$ is transformed into†

$$a \to a' = a \cos\theta/2 + b \sin\theta/2$$

$$b \to b' = -a\sin\theta/2 + b\cos\theta/2$$

or

$$\begin{pmatrix} a \\ b \end{pmatrix} \to \begin{pmatrix} a' \\ b' \end{pmatrix} = \begin{pmatrix} \cos\theta/2 & \sin\theta/2 \\ -\sin\theta/2 & \cos\theta/2 \end{pmatrix} \begin{pmatrix} a \\ b \end{pmatrix}. \tag{3-15}$$

---

† This formula is a direct translation from spin ½ to isospin ½. For the spin case, see, for example, R. P. Feynman, R. B. Leighton and M. Sands, *The Feynman Lectures on Physics* vol. 3, Addison-Wesley 1965, Chapter 6.

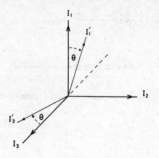

Figure 3.5

This may be written, using (6), as

$$\begin{pmatrix} a \\ b \end{pmatrix} \rightarrow \begin{pmatrix} a' \\ b' \end{pmatrix} = \left[\begin{pmatrix} 1 & 0 \\ 0 & 1 \end{pmatrix} \cos\theta/2 + i\tau_2 \sin\theta/2 \right] \begin{pmatrix} a \\ b \end{pmatrix},$$

and since $(\tau_2)^2 = \begin{pmatrix} 1 & 0 \\ 0 & 1 \end{pmatrix}$, this in turn is

$$\begin{pmatrix} a \\ b \end{pmatrix} \rightarrow \begin{pmatrix} a' \\ b' \end{pmatrix} = \exp(i\tau_2\theta/2) \begin{pmatrix} a \\ b \end{pmatrix} = \exp(iI_2\theta) \begin{pmatrix} a \\ b \end{pmatrix} \quad (3\text{-}16)$$

EXERCISE Prove (16), by expanding the exponential.

Equation (16) represents a rotation about the "2" axis in isospin space, just as equation (2-21) represents a rotation about the z axis in ordinary space. Ignoring the factor $\hbar$, we may make the associations

$\left. \begin{array}{l} L_x, L_y, L_z \\ \text{in ordinary space} \end{array} \right\}$ correspond to $\left\{ \begin{array}{l} I_1, I_2, I_3 \\ \text{in isospin space.} \end{array} \right.$

I_1, I_2 and I_3 (or equivalently I_+, I_- and I_3) are the *generators* of the group of rotations in isospin space. This group is mathematically identical to the group of rotations in ordinary space, i.e. SO_3 or SU_2.

What is the physical meaning of transformations like (15) and (16)? These transformations change the values of a and b; and from (5) a is the amplitude that the nucleon is a proton, and b the amplitude that it is a neutron. So by changing these amplitudes, the isospin rotation effectively *mixes up* proton and neutron states. Since the strong interactions are charge independent, they do not distinguish the proton and neutron, and are therefore *invariant* under rotations in isospin space.

As far as conservation laws are concerned, applying the reasoning in Chapter 2 leads to the conclusion

$$\begin{cases} \text{strong interactions conserve isospin} \\ \text{electromagnetic interactions do not conserve isospin} \end{cases} \quad (3\text{-}17)$$

If we split the Hamiltonian (energy) operator for the nucleon system into strong and electromagnetic parts

$$H = H_{st} + H_{em}.$$

this implies that (in analogy with equation (2-19))

$$[I, H_{st}] = IH_{st} - H_{st}I = 0 \quad (3\text{-}18)$$

$$[I, H_{em}] = IH_{em} - H_{em}I \neq 0 \quad (3\text{-}19)$$

Thus Table 2-1 may be extended by adding the entry of Table 2, with the crucial proviso that unlike the other symmetries, isospin symmetry is a symmetry of the strong interactions *only*.[†] Since it is the electromagnetic interactions which distinguish the proton and neutron states, then we have another illustration of the association discussed at the beginning of Chapter 2

symmetry ↔ *indistinguishability*

breaking of symmetry ↔ *distinguishability*

Table 3.2

Indistinguishability of the strong interactions of	implies invariance of their strong interactions under	and conservation of
proton and neutron	rotations in isospin space	isospin

---

† We shall see in Chapter 5 that the weak interactions, like the electromagnetic ones, do not conserve isospin.

3.7 ISOSPIN CONSERVATION IN STRONG INTERACTIONS

Let us continue to argue by considering the analogy with ordinary angular momentum. In the last chapter it was stated (see (2-19)) that **L** is conserved. As the reader well knows, however, when the particles have spin, the conserved quantum number is **J** = **L** + **S**, the total angular momentum. The consequence of this is that[†]

J is a "good" quantum number: the states of a system may be labelled by **J** and by the eigenvalues of J_z. There are $2J + 1$ of these, and they are all degenerate. (3-20)

Now let us assume that not only the strong interactions of protons and neutrons, but *all* strong interactions conserve isospin. The analogous consequence is that

As far as the strong interactions are concerned, **I** is a "good" quantum number: the states of a system (elementary particles with strong interactions) may be labelled by **I** and by the eigenvalues of I_3. There are $2I + 1$ of these, and they are degenerate. (3-21)

This rule would only be exact if there were only strong interactions in nature. However, although there are other interactions, the strong interactions are the dominant ones, so a symmetry of the strong interactions is an *approximate* symmetry of nature. We have, in place of (21)

I is a "good" quantum number for strong interactions. The states of a system of strongly interacting particles (hadrons) may be labelled by **I** and by the eigenvalues of I_3. There are $2I + 1$ of these, and they are *approximately degenerate*. (3-22)

I, like **J**, is a vector quantum number, and we shall use the notation $I = |\mathbf{I}|$. At the risk of tedious repetition, let me repeat that the nucleon N is an example of (22). It is a system with $I = \frac{1}{2}$, and the two components with $I_3 = \pm\frac{1}{2}$ are p and n, which are approximately degenerate.

3.8 IMPLICATIONS OF CHARGE INDEPENDENCE FOR PIONS

Since pions are "traditionally" (see Chapter 1) the agents of nuclear forces (strong interactions), it is natural to enquire what consequences we can deduce about them from the charge symmetry and charge independence of these forces.

[†] See, e.g., R. M. Eisberg, *Fundamentals of Modern Physics,* Wiley, 1961, section 11.5.

ISOSPIN 53

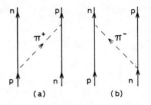

Figure 3.6

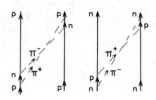

Figure 3.7

As explained in Chapter 1 the *p-n* force is caused by the exchange of a charged pion — a π^+ one way, or a π^- the other way, as in Figure 6. The range R of the force is given by

$$R_{pn} \propto \frac{1}{m_{\pi^+}},$$

and $m_{\pi^+} = m_{\pi^-}$. If there were only π^+ and π^-, the *p-p* and *n-n* forces would have to be implemented by the exchange of both a π^+ and a π^-, so as to result in no charge exchange (Figure 7). This force would have range

$$R_{nn} = R_{pp} \propto \frac{1}{2m_{\pi^+}} = \tfrac{1}{2} R_{pn},$$

and would be stronger than the *p-n* force, since two pions can "glue" more strongly than one.

But the observation of charge independence is precisely the observation that the *p-n* force is equal to the *p-p* and *n-n* forces, both in magnitude and in range. This implies that there must be a *third* pion with zero charge π^0, *with the same mass as* π^+ *and* π^-. In this case, the *p-p* and *n-n* forces are due to exchanges of a single pion π^0, as in Figure 8, and the range

$$R_{nn} = R_{pp} = R_{pn} \propto \frac{1}{m_{\pi^0}} = \frac{1}{m_{\pi^+}}$$

Note that charge symmetry is not a strong enough condition to deduce the existence of π^0; we need charge independence.

It would be difficult to overstress the importance of this deduction from the charge independence of nuclear forces:—

There must be three pions, with charges +1, 0, −1, which have identical strong interactions with nucleons, and the same mass.

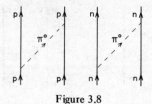

Figure 3.8

Experimentally, the masses of the pions are $m(\pi^+) = m(\pi^-) = 139.6$ MeV; $m(\pi^0) = 135.0$ MeV. It is generally believed that the fact that $m(\pi^+) \neq m(\pi^0)$ is due to the electromagnetic interactions of $\pi^\pm$ and π^0, which have the effect of making the charged pions slightly heavier; and that, exactly as for the nucleon, in the fictitious world where the electromagnetic interactions are turned off, π^+, π^- and π^0 are identical.

These deductions from charge independence were first made by Kemmer.

3.9 PIONS AND ISOSPIN

Let us cast the result of the last section into a different form by discussing it solely in terms of isospin. This turns out to be more concise, and admits of a clear generalisation to particles other than pions, and an illustration of statement (22).

We shall begin by considering the diagram in Figure 6(a). Call the points where the pion and nucleon lines meet, "vertices". In Figure 6(a), then, there are two vertices. At the first of these we have the "reaction"

$$p \to n + \pi^+$$

and at the second

$$n + \pi^+ \to p.$$

These reactions are intimately related to

$$p + \bar{n} \to \pi^+ \tag{3-23}$$

and

$$n + \bar{p} \to \pi^- \tag{3-24}$$

where $\bar{n}$ and $\bar{p}$ denote antiproton and antineutron, and we have used the rule that if we take a particle from one side of a reaction, we replace it by its antiparticle on the other side. By the same rule, the reactions at the vertices in Figure 7

$$p \to p + \pi^0; \quad n \to n + \pi^0$$

become

$$p + \bar{p} \to \pi^0 \tag{3-25}$$

$$n + \bar{n} \to \pi^0. \tag{3-26}$$

Since, as far as strong interactions are concerned, p and n are identical, reactions (23) to (26) may be written

$$N + \bar{N} \to \pi \tag{3-27}$$

where N is the nucleon and π is the pion.

Now focus attention on (27). N is a particle with isospin $I = \frac{1}{2}$, and $\bar{N}$ also has $I = \frac{1}{2}$. What is the isospin of π? We have already seen that isospin was conceived as being *mathematically* identical to spin. So, mathematically, our question is identical to the question: in the reaction $a + b \to c$, if a and b both have spin $\frac{1}{2}$, what is the spin of c? Assuming *no orbital angular momentum is involved* (since there is no analogous "orbital isospin"), the answer is 0 or 1. The general rule, known as the *vector addition rule*, is[†]

$$s_c = |s_a - s_b|, |s_a - s_b| + 1, \ldots, s_a + s_b.$$

If $s_a = s_b = \frac{1}{2}$, this gives $s_c = 0, 1$. So, returning to reaction (27), we see that if $I_N = I_{\bar{N}} = \frac{1}{2}$, then $I_\pi = 0, 1$. Or, in general, if $a + b \to c$, then

$$I_c = |I_a - I_b|, |I_a - I_b| + 1, \ldots, I_a + I_b \tag{3-28}$$

Now if $I_\pi = 0$ then $2I_\pi + 1 = 1$ and there is one pion. If $I_\pi = 1$, there are $2I_\pi + 1 = 3$ pions. It turns out that, between these possibilities, nature has chosen $I_\pi = 1$.

We now see how statement (22) above applies to the pion. The pion is a particle with $I = 1$, and so comprises $2I + 1 = 3$ states with $I_3 = +1, 0, -1$. These are π^+, π^0, π^-. They are identical apart from their charge, and are *approximately degenerate*.

The fact that nature has chosen three pions and not one is due to the nature of the Yukawa force. If the force were as in Figure 9 there would only be one pion, π^0. But the *p-n* force is not like this, it is an exchange force as in Figure 6, so

---

[†] See, for example, R. B. Leighton, *Principles of Modern Physics*, McGraw-Hill 1959, section 7; or Eisberg, op.cit., p.431.

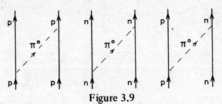

Figure 3.9

there must be at least one other pion, with a charge. In other words, there must be three altogether. The condition of an exchange force has nothing to do with isospin; it is an additional constraint which makes the number of pions three instead of one.

Finally, let us summarise this by writing down the analogue of (3) for the pion

$$\text{pion } \pi \text{ isospin } I = 1 \begin{cases} I_3 = +1 \ \pi^+ \\ I_3 = 0 \ \pi^0 \\ I_3 = -1 \ \pi^- \end{cases} \tag{3-29}$$

And, to match (4), we have

$$\pi: Q = I_3 \tag{3-30}$$

3.10 RAISING AND LOWERING OPERATORS FOR THE PION

We should like to know, for the pion isotriplet, the analogue of (11) to (14) for the nucleon isodoublet. We expect that

$$I_- \pi^+ = a\pi^0$$

where a is a numerical coefficient, since I_- is defined to be the operator which lowers the value of I_3 by 1. The question is, what is the value of a? The answer may be obtained by assuming, in the spirit of equations (23) to (26), that *the pion is a nucleon-antinucleon bound state.* In fact it is *not*, it is just as elementary as p and n, but for the present purposes this does not matter. The nucleon and antinucleon doublets are

$$N = \begin{pmatrix} p \\ n \end{pmatrix}, \quad \bar{N} = \begin{pmatrix} -\bar{n} \\ \bar{p} \end{pmatrix}. \tag{3-31}$$

Note that it is $\bar{n}$ which has isospin up ($I_3 = \frac{1}{2}$), because it has the higher charge. The minus sign is put in to make the phases consistent. If $\pi = (N\bar{N})$, then it is obvious that

$$\pi^+ = -(p\bar{n}), \quad \pi^- = (n\bar{p}) \tag{3-32}$$

ISOSPIN

to get the charges right. What, then, is π^0? It could be $p\bar{p}$ or $n\bar{n}$, so we write as the combination

$$\pi^0 = \frac{1}{\sqrt{2}}(p\bar{p} - n\bar{n}) \qquad (3\text{-}33)$$

where the factor $1/\sqrt{2}$ is to normalise the π^0 wave function. (The minus signs in (32) and (33) are simply consequences of the one in (31).) From (13) and (14) we have

$$I_- p = n, \; I_- n = 0. \qquad (3\text{-}34)$$

We also have, from (31)

$$I_- \bar{n} = -\bar{p}, \; I_- \bar{p} = 0. \qquad (3\text{-}35)$$

Equations (32), (34) and (35) now give

$$I_- \pi^+ = -I_-(p\bar{n})$$

$$= -(I_- p)\bar{n} - p(I_- \bar{n})$$

$$= -n\bar{n} + p\bar{p}$$

$$= \sqrt{2}\pi^0$$

where in the last step we have used (33). This answers our original question: $a = \sqrt{2}$. In the same way it is easy to prove that

$$I_- \pi^+ = \sqrt{2}\pi^0, \; I_- \pi^0 = \sqrt{2}\pi^-, \; I_- \pi^- = 0$$

$$I_+ \pi^- = \sqrt{2}\pi^0, \; I_+ \pi^0 = \sqrt{2}\pi^+, \; I_+ \pi^+ = 0. \qquad (3\text{-}36)$$

These operations are shown diagramatically in Figure 10.

EXERCISE Prove equation (36).

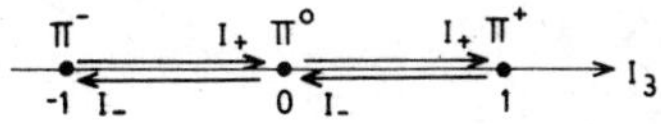

Figure 3.10 I_+ and I_- operating on the pion states π^+, π^0 and π^-

3.11 COMPOSITION OF ISOSPINS

There is a general formula of which equations (11) to (14) and (36) are special cases. It is

$$I_+(I, I_3) = \sqrt{(I - I_3)(I + I_3 + 1)}(I, I_3 + 1) \tag{3-37}$$

$$I_-(I, I_3) = \sqrt{(I + I_3)(I - I_3 + 1)}(I, I_3 - 1) \tag{3-38}$$

where (I, I_3) is the state with isospin I and third component I_3.

EXERCISE   Check that (37) and (38) give (11) − (14) and (36).

In fact, we could have performed the calculation on the pions backwards, starting with (36), which would follow from (37) and (38), and finishing up with (32) and (33). In that case the problem would have been: from the two $I = \frac{1}{2}$ multiplates (p, n) and $(-\bar{n}, \bar{p})$, what are the $I_3 = 1, 0, -1$ states of the $I = 1$ multiplet obtained from them? The answer would have been (using the notation (I, I_3))

(π^+): $(1, 1) = -(p\bar{n})$

(π^0): $(1, 0) = \dfrac{1}{\sqrt{2}}(p\bar{p} - n\bar{n})$

(π^-): $(1, -1) = (n\bar{p})$ \hfill (3-39)

The importance of this is that the same method may be used to construct higher isospin states. For example, consider the states of the pion-nucleon system. If isospin is a good quantum for their production, they must, by the vector addition rule, have $I = \frac{3}{2}$ or $I = \frac{1}{2}$. We should then want to know how the states $(\frac{3}{2}, \frac{3}{2})$, $(\frac{3}{2}, \frac{1}{2})$, etc., are expressed in terms of π^+, π^0, π^- and p, n. To find out we proceed exactly as above, using equations (11) − (14) and (36) − (38), and get, for example

$$(\tfrac{3}{2}, \tfrac{3}{2}) = \pi^+ p \tag{3-40}$$

$$(\tfrac{3}{2}, \tfrac{1}{2}) = \sqrt{\tfrac{2}{3}}(\pi^0 p) + \sqrt{\tfrac{1}{3}}(\pi^+ n) \tag{3-41}$$

$$(\tfrac{1}{2}, \tfrac{1}{2}) = \sqrt{\tfrac{1}{3}}(\pi^0 p) - \sqrt{\tfrac{2}{3}}(\pi^+ n). \tag{3-42}$$

The complete derivation is given in the appendix at the end of this chapter, which also gives the (2π) states with $I = 2, 1, 0$.

ISOSPIN

3.12 GENERALISED PAULI EXCLUSION PRINCIPLE: THE DEUTERON

The deuteron d, the nucleus of deuterium, is a proton-neutron bound state, bound by strong interactions. According to (22), then, it must have a definite isospin. What is it?

Within the context of strong interactions, p and n are identical particles with $s = \frac{1}{2}, I = \frac{1}{2}$. According to Pauli's exclusion principle, no two identical fermions at the same place (strictly, whose wave functions overlap) may be in the same state.† There is a more sophisticated way of stating this, and that is that the wave function is antisymmetric under exchange of particle coordinates. Let me refresh the reader's memory by proving the equivalence of these statements. For simplicity, for the moment forget spin, and consider two identical fermions a and b. Remembering that in quantum theory the wave function specifies the *amplitude* for a state, let us say that the amplitude for a to be at the point x is $\psi_a(x)$ and the amplitude for b to be at y is $\psi_b(y)$. Then the amplitude for a to be at x and b to be at y is

$$\psi_a(x)\psi_b(y).$$

Similarly, the amplitude for a to be at y and b to be at x is

$$\psi_a(y)\psi_b(x).$$

Now since the fermions are identical, and therefore *indistinguishable*, we can only ask for the amplitude that there is one particle at x and one at y, and this amplitude will clearly contain both those above, in equal portions. The exclusion principle tells us that we take the antisymmetric combination, so the total amplitude is

$$A = \frac{1}{\sqrt{2}} \left[\psi_a(x)\psi_b(y) - \psi_a(y)\psi_b(x)\right].$$

This has the desired antisymmetry property that under the interchange $x \leftrightarrow y$, $A \rightarrow -A$. But A also has the property that as $x \rightarrow y$, $A \rightarrow 0$, so the amplitude for finding the fermions at the same point is zero. This was the original statement, and completes the proof.

To include spin, we merely enlarge the exclusion principle to state that under exchange of space *and* spin coordinates, the amplitude A should be antisymmetric. The spin wave functions are u and v, the amplitudes for the particles to have spin up or spin down. The total wave functions appear in Table 3 where it will be seen that the wave function for total spin $S = 1$ is symmetric in spin (and therefore

---
† For a fuller account, see, e.g., Eisberg, op.cit., Chapter 12.

Table 3.3

The wave functions for identical spin ½ particles a and b, totally antisymmetric under exchange of space (x, y) and spin (u, v) coordinates. The arrows represent the spin up and down states, u and v

State	Wave function	S_z	S
↑↓	$\frac{1}{2}[\psi_a(x)\psi_b(y) + \psi_a(y)\psi_b(x)][u_a v_b - v_a u_b]$	0	0
↑↑	$\frac{1}{\sqrt{2}}[\psi_a(x)\psi_b(y) - \psi_a(y)\psi_b(x)]u_a u_b$	1	
↑↓	$\frac{1}{2}[\psi_a(x)\psi_b(y) - \psi_a(y)\psi_b(x)][u_a v_b + v_a u_b]$	0	1
↓↓	$\frac{1}{\sqrt{2}}[\psi_a(x)\psi_b(y) - \psi_a(y)\psi_b(x)]v_a v_b$	−1	

antisymmetric in space) coordinates, and that for $S = 0$ is antisymmetric in spin (and symmetric in space) coordinates.†

To include isospin, we enlarge the exclusion principle yet further to a "generalised exclusion principle" which states that the total wave function for two identical fermions should be antisymmetric under the simultaneous interchange of space, spin and isospin coordinates. The wave function is written

$$\Psi = \psi_{\text{space}} \psi_{\text{spin}} \psi_{\text{isospin}} \qquad (3\text{-}43)$$

where

$$\psi_{\text{space}} = \sqrt{\tfrac{1}{2}}[\psi_a(x)\psi_b(y) \pm \psi_a(y)\psi_b(x)] \qquad (3\text{-}44)$$

$$\psi_{\text{spin}} = \text{one of} \begin{cases} \sqrt{\tfrac{1}{2}}(u_a v_b - v_a u_b) \\ u_a u_b \\ \sqrt{\tfrac{1}{2}}(u_a v_b + v_a u_b) \\ v_a v_b \end{cases} \qquad (3\text{-}45)$$

and, analogously

---

† See, e.g., Eisberg, op.cit., p.370.

ISOSPIN

$$\psi_{\text{isospin}} = \text{one of} \begin{cases} \sqrt{\tfrac{1}{2}}(p_a n_b - n_a p_b) \\ p_a p_b \\ \sqrt{\tfrac{1}{2}}(p_a n_b + n_a p_b) \\ n_a n_b \end{cases} \quad (3\text{-}46)$$

What this means is that the wave function Ψ represents the amplitude for two nucleons, a and b, to be at the points x and y, with different combinations of spin up and spin down (u and v) and of isospin up and isospin down (p and n). These isospin wave functions are noted in Table 4, which is a direct "translation" of the spin wave function of Table 3 to isospin wave functions.

Now if the total wave function (43) is antisymmetric under the simultaneous exchange of space, spin and isospin coordinates, then either all three or only one of the functions (44) to (46) must be antisymmetric. The deuteron is mainly in a $^3S_1$ state, i.e. $L = 0, S = 1$. There is a small admixture of $^3D_1$, i.e. $L = 2, S = 1$. The space symmetry ($x \leftrightarrow y$) is given by $(-1)^L = +1$ for $L = 0$ or 2. So ψ_{space} is symmetric.

Since $S = 1$, from Table 3 ψ_{spin} is also symmetric. So ψ_{isospin} must be antisymmetric, and the complete wave function is

$$\Psi = \frac{1}{2\sqrt{2}} \underbrace{[\psi_a(x)\psi_b(y) + \psi_a(y)\psi_b(x)]}_{\substack{\text{symmetric under} \\ x \leftrightarrow y}} \underbrace{[u_a v_b + v_a u_b]}_{\substack{\text{symmetric} \\ \text{under } u \leftrightarrow v}} \underbrace{[p_a n_b - n_a p_b]}_{\substack{\text{antisymmetric} \\ \text{under } p \leftrightarrow n}}$$

Table 3.4
The wave functions for two identical nucleons a and b, symmetric and antisymmetric under exchange of isospin coordinates (p, n)

State	Isospin wave function	I_3	I
$p + n$	$\frac{1}{\sqrt{2}}(p_a n_b - n_a p_b)$	0	0
$p + p$	$p_a p_b$	1	
$p + n$	$\frac{1}{\sqrt{2}}(p_a n_b + n_a p_b)$	0	1
$n + n$	$n_a n_b$	-1	

In other words, from Table 4, d has $I = 0$. This is an important prediction, for it says that d is an isosinglet and therefore *has no "partners" with difference charge*, e.g. (pp) or (nn). This is indeed what is found experimentally. There is no pp or nn state similar to d. In analogy, then, to (3) and (28) we have

$$\text{deuteron } d \text{ isospin } I = 0 \quad\text{———}\quad I_3 = 0 \; d \tag{3-47}$$

and, since d has $B = 2$, $Q = 1$ and $I_3 = 0$, we have the relation

$$d: Q = I_3 + 1. \tag{3-48}$$

3.13 TESTS FOR ISOSPIN CONSERVATION IN STRONG INTERACTIONS

First, consider the reactions

(1) $\quad n + p \to \pi^0 + d$
$\quad I \quad\; ½ \quad ½ \quad 1 \quad 0$
$\quad I_3 \; -½ \quad ½ \quad 0 \quad 0$

(2) $\quad p + p \to \pi^+ + d$
$\quad I \quad\; ½ \quad ½ \quad 1 \quad 0$
$\quad I_3 \;\; ½ \quad ½ \quad 1 \quad 0$

where the values of I and I_3 have been noted for each particle. In both reactions the right hand side has $I = 1$. So, if the reactions conserve I, the left hand sides must also have $I = 1$. The left hand side of (2) is in a pure $I = 1$ state, since $I_3 = ½ + ½ = 1$. In reaction (1), however, $n + p$ has $I_3 = ½ - ½ = 0$, and this is equally likely to belong to $I = 1$ or $I = 0$. $n + p$ will therefore have $I = 1$ for half the time only, whereas $p + p$ always has $I = 1$. So the ratio of the cross sections should be

$$R = \frac{\sigma(n + p \to \pi^0 + d)}{\sigma(p + p \to \pi^+ + d)} = ½$$

since only when $n + p$ is in an $I = 1$ state, will it react to give $\pi^0 + d$. This is Yang's test for isospin conservation in strong interactions. Experimentally it is found that $R = ½$ to within 10%.

As a further test consider the reactions

$p + d \to \pi^0 + He^3$

$p + d \to \pi^+ + H^3$.

ISOSPIN

Since d has $I = 0$, the initial state in these reactions has $(I, I_3) = (\tfrac{1}{2}, \tfrac{1}{2})$. Conservation of I demands that the final state have the same values. Now π has $I = 1$, and (He^3, H^3) form an isodoublet like (p, n). The $(\tfrac{1}{2}, \tfrac{1}{2})$ state of (π^+, π^0, π^-) and (He^3, H^3) is obtained from (42) by substituting $(p, n) \to (He^3, H^3)$:

$$(\tfrac{1}{2}, \tfrac{1}{2}) = \sqrt{\tfrac{1}{3}}(\pi^0 + He^3) - \sqrt{\tfrac{2}{3}}(\pi^+ + H^3).$$

It follows that

$$\frac{\text{amplitude } (p + d \to \pi^+ + H^3)}{\text{amplitude } (p + d \to \pi^0 + He^3)} = -\sqrt{2},$$

and therefore, since cross section $\sigma \propto |\text{amplitude}|^2$,

$$\frac{\sigma(p + d \to \pi^+ + H^3)}{\sigma(p + d \to \pi^0 + He^3)} = 2.$$

Experimentally, this ratio has been found to be 2, to an accuracy slightly better than 10%.

3.14 G-PARITY

Charge conjugation C has the property that only neutral states are C-eigenstates. Using isospin a generalisation of C, called G conjugation may be defined, which has the property that charged as well as neutral states are eigenstates of G. The eigenvalue of G is called the *G-parity* of the state. G is defined by

$$G = \exp(i\pi I_2)C$$

where $\exp(i\pi I_2)$ is the operator for rotation through 180° about the second isospin axis. Define the pion field as

$$\pi^+ = \frac{1}{\sqrt{2}}(\pi_1 + i\pi_2)$$

$$\pi^- = \frac{1}{\sqrt{2}}(\pi_1 - i\pi_2)$$

$$\pi^0 = \pi_3$$

so that

$$C\pi^\pm = \pi^\mp, \quad C\pi^0 = \pi^0$$

implies that

$$\begin{pmatrix} \pi_1 \\ \pi_2 \\ \pi_3 \end{pmatrix} \xrightarrow{C} \begin{pmatrix} \pi_1 \\ -\pi_2 \\ \pi_3 \end{pmatrix}.$$

In addition

$$\begin{pmatrix} \pi_1 \\ \pi_2 \\ \pi_3 \end{pmatrix} \xrightarrow{\exp(i\pi I_2)} \begin{pmatrix} -\pi_1 \\ \pi_2 \\ -\pi_3 \end{pmatrix},$$

so that

$$\begin{pmatrix} \pi_1 \\ \pi_2 \\ \pi_3 \end{pmatrix} \xrightarrow{G} \begin{pmatrix} -\pi_1 \\ -\pi_2 \\ -\pi_3 \end{pmatrix}$$

and the pion is in an eigenstate of G with eigenvalue (G-parity) -1:

$$G\pi^+ = -\pi^+, \quad G\pi^- = -\pi^-, \quad G\pi^0 = -\pi^0.$$

Clearly G is a good quantum number if I and C are, that is, in strong interactions, since electromagnetic interactions violate I. This means that since an odd number of pions has $G = -1$ and an even number $G = 1$, strong transitions between an odd and an even number of pions are forbidden.

G-parity was introduced by Michel who called it isotopic parity. C is reflection in a plane — in this case the (13) plane. $\exp(i\pi I_2)$ is rotation about the "2" axis (perpendicular to the plane) through π. This is analogous to the definition of space reflection (cf. section 8.2).

3.15 SUMMARY

In summarising the results of the present chapter, let us cast them in a slightly different form to emphasize the new way we have learned of looking at the strong interactions.

1) The proton, neutron, the three pions and the deuteron exist in *charge families:* that is, families of almost identical particles, whose electric charges

ISOSPIN 65

differ in steps of one. If there are $2I + 1$ particles in the family, we say we have a particle of isospin I. Thus the nucleon N has $I = \frac{1}{2}(p, n)$, the pion π has $I = 1$ (π^+, π^0, π^-), the deuteron d has $I = 0$ (d).

2) Isospin is conserved in the strong interactions between these particles. Since I is a vector in isospin space, we use the *vector addition rule* to add isospins.

FURTHER READING

Kemmer, N., J. C. Polkinghorne and D. L. Pursey, "Invariance in Elementary Particle Physics", *Reports on Progress in Physics*, **22**, 368 (1959).
Marshak, R. E. and E. C. G. Sudarshan, *Introduction to Elementary Particle Physics*, Interscience, 1961, Chapter 5.
Nishijima, K., *Fundamental Particles*, Benjamin, 1963, Chapter 3.

Appendix

COMPOSITION OF ISOSPINS

Consider an isomultiplet with isospin I, and $I_3 = I, I - 1, \ldots, -I$. Label the $2I + 1$ states (I, I_3). Define the raising and lowering operators I_+ and I_- by the relations (cf. (3-37) and 3-38))

$$I_+(I, I_3) = \sqrt{(I - I_3)(I + I_3 + 1)}(I, I_3 + 1) \tag{3A-1}$$

$$I_-(I, I_3) = \sqrt{(I + I_3)(I - I_3 + 1)}(I, I_3 - 1). \tag{3A-2}$$

Applying the above definitions to the nucleon N with $I = \frac{1}{2}$; $(\frac{1}{2}, \frac{1}{2}) = p$, $(\frac{1}{2}, -\frac{1}{2}) = n$, gives

$$I_+ n = p \tag{3A-3}$$

$$I_- p = n \tag{3A-4}$$

$$I_+ p = I_- n = 0 \tag{3A-5}$$

$$I_3 p = \tfrac{1}{2} p, \; I_3 n = -\tfrac{1}{2} n. \tag{3A-6}$$

Note that the coefficients in (1) and (2) have been chosen to give (5), which expresses the fact that there are no particles in the multiplet with a higher value of I_3 than p, or a lower one than n. Similarly applying (1) and (2) to the pion π ($I = 1$: $(1, 1) = \pi^+$, $(1, 0) = \pi^0$, $(1, -1) = \pi^-$) gives

$$I_+ \pi^- = \sqrt{2} \pi^0 \tag{3A-7}$$

$$I_+ \pi^0 = \sqrt{2} \pi^+ \tag{3A-8}$$

$$I_- \pi^+ = \sqrt{2} \pi^0 \tag{3A-9}$$

$$I_- \pi^0 = \sqrt{2} \pi^- \tag{3A-10}$$

$$I_+ \pi^+ = I_- \pi^- = 0 \tag{3A-11}$$

$$I_3 \pi^+ = \pi^+ \tag{3A-12}$$

$$I_3 \pi^0 = 0 \tag{3A-13}$$

$$I_3 \pi^- = -\pi^-. \tag{3A-14}$$

ISOSPIN

Now consider a system of two particles A and B with isospins I_A and I_B. Using the vector addition rule, the resultant isospin is

$$I = I_A + I_B, I_A + I_B - 1, \ldots, |I_A - I_B|. \tag{3A-15}$$

The question we are interested in is; for each possible isospin I in (15), how are the constituent states (I, I_3) made up out of the states of A and B? To work this out, note that

$$I_\pm = I_\pm^A + I_\pm^B. \tag{3A-16}$$

πN system

For definiteness consider the πN system. From the vector addition rule $I = 3/2$ or $1/2$, so there are 6 states in all: $(3/2, 3/2)$, $(3/2, 1/2)$, $(3/2, -1/2)$, $(3/2, -3/2)$; and $(1/2, 1/2)$, $(1/2, -1/2)$. These must be expressed in terms of pion and nucleon states. To begin with, it is clear that the "highest" state must be

$$(3/2, 3/2) = \pi^+ p \tag{3A-17}$$

since it is the only one with $I_3 = 3/2$. Now apply I_- to both sides. The lhs gives from (2)

$$I_-(3/2, 3/2) = \sqrt{3}(3/2, 1/2) \tag{3A-18}$$

and the rhs, from (16), is

$$I_-(\pi^+ p) = I_-^\pi (\pi^+ p) + I_-^N (\pi^+ p)$$

$$= (I_- \pi^+) p + \pi^+ (I_- p)$$

$$= \sqrt{2}(\pi^0 p) + (\pi^+ n) \tag{3A-19}$$

where (4) and (9) have been used. (17)–(19) now give

$$(3/2, 1/2) = \sqrt{2/3}(\pi^0 p) + \sqrt{1/3}(\pi^+ n). \tag{3A-20}$$

Applying I_- to (20) gives

$$I_-(3/2, 1/2) = 2(3/2, -1/2) \tag{3A-21}$$

and

$$I_-[\sqrt{2/3}(\pi^0 p) + \sqrt{1/3}(\pi^+ n)]$$

$$= \sqrt{2/3}[(I_-\pi^0)p + \pi^0(I_-p)] + \sqrt{1/3}[(I_-\pi^+)n + \pi^+(I_-n)]$$

$$= \sqrt{2/3}[\sqrt{2}(\pi^-p) + (\pi^0 n)] + \sqrt{1/3}[\sqrt{2}(\pi^0 n) + 0]$$

$$= \frac{2}{\sqrt{3}}(\pi^-p) + \frac{2\sqrt{2}}{\sqrt{3}}(\pi^0 n). \tag{3A-22}$$

Equations (20)–(22) give

$$(3/2, -1/2) = \sqrt{1/3}(\pi^-p) + \sqrt{2/3}(\pi^0 n). \tag{3A-23}$$

There is no need to apply I_- to (23) since it is obvious that

$$(3/2, -3/2) = \pi^-n, \tag{3A-24}$$

the only state with $I_3 = -3/2$.

The remaining two states $(1/2, 1/2)$ and $(1/2, -1/2)$ are obtained from the condition that they must be orthogonal to $(3/2, 1/2)$ and $(3/2, -1/2)$ respectively. This gives

$$(1/2, 1/2) = \sqrt{2/3}(\pi^+n) - \sqrt{1/3}(\pi^0 p)$$

$$(1/2, -1/2) = \sqrt{1/3}(\pi^0\pi) - \sqrt{2/3}(\pi^-p) \tag{3A-25}$$

in addition to the states obtained already

$$(3/2, 3/2) = \pi^+p$$

$$(3/2, 1/2) = \sqrt{2/3}(\pi^0 p) + \sqrt{1/3}(\pi^+n)$$

$$(3/2, -1/2) = \sqrt{1/3}(\pi^-p) + \sqrt{2/3}(\pi^0 n)$$

$$(3/2, -3/2) = \pi^-n. \tag{3A-26}$$

(It is easy to verify from (25) that $I_+(1/2, 1/2) = I_-(1/2, -1/2) = 0$ and $I_+(1/2, -1/2) = (1/2, 1/2)$.)

(25) and (26) may be inverted to give

$$(\pi^+p) = (3/2, 3/2)$$

$$(\pi^0 p) = \sqrt{2/3}(3/2, 1/2) - \sqrt{1/3}(1/2, 1/2)$$

$(\pi^+ n) = \sqrt{1/3}(3/2, 1/2) + \sqrt{2/3}(1/2, 1/2)$

$(\pi^- p) = \sqrt{1/3}(3/2, -1/2) - \sqrt{2/3}(1/2, -1/2)$

$(\pi^0 n) = \sqrt{2/3}(3/2, -1/2) + \sqrt{1/3}(1/2, -1/2)$

$(\pi^- n) = (3/2, -3/2).$ \hfill (3A-27)

The easiest way to present the results for any isospin is in tables of the numerical coefficients (called *Clebsch-Gordon coefficients*), which are set out below. They should be self-explanatory, except to remark that the notation (J, M), (j, m) instead of (I, I_3) is used.

As a final example of isospin composition, the 1×1 table gives for the $(I, I_3 = 0)$ states of two pions

$$(2, 0) = \frac{1}{\sqrt{6}}(\pi^-\pi^+ + \pi^+\pi^- + 2\pi^0\pi^0)$$

$$(1, 0) = \frac{1}{\sqrt{2}}(\pi^-\pi^+ - \pi^+\pi^-)$$

$$(0, 0) = \frac{1}{\sqrt{3}}(\pi^+\pi^- + \pi^-\pi^+ - \pi^0\pi^0) \hfill (3A\text{-}28)$$

It is worth noting that the $(2, 0)$ and $(0, 0)$ states are symmetric under $\pi^+ \leftrightarrow \pi^-$, whereas the $(1, 0)$ state is antisymmetric.

½ × ½

m_1	m_2	J M	1 1	1 0	0 0	1 −1
1/2	1/2		1			
1/2 −1/2	−1/2 1/2			$\sqrt{1/2}$ $\sqrt{1/2}$	$\sqrt{1/2}$ $-\sqrt{1/2}$	
−1/2	−1/2					1

$1 \times 1/2$

m_1	m_2	J M	3/2 3/2	3/2 1/2	1/2 1/2	3/2 -1/2	1/2 -1/2	3/2 -3/2
1	1/2		1					
1	-1/2			$\sqrt{1/3}$	$\sqrt{2/3}$			
0	1/2			$\sqrt{2/3}$	$-\sqrt{1/3}$			
0	-1/2					$\sqrt{2/3}$	$\sqrt{1/3}$	
-1	1/2					$\sqrt{1/3}$	$-\sqrt{2/3}$	
-1	-1/2							1

$3/2 \times 1/2$

m_1	m_2	J M	2 2	2 1	1 1	2 0	1 0	2 -1	1 -1	2 -2
3/2	1/2		1							
3/2	-1/2			1/2	$\sqrt{3/2}$					
1/2	1/2			$\sqrt{3/2}$	-1/2					
1/2	-1/2					$\sqrt{1/2}$	$\sqrt{1/2}$			
-1/2	1/2					$\sqrt{1/2}$	$-\sqrt{1/2}$			
-1/2	-1/2							$\sqrt{3/2}$	1/2	
-3/2	1/2							1/2	$-\sqrt{3/2}$	
-3/2	-1/2									1

ISOSPIN

$2 \times \frac{1}{2}$

m_1	m_2		J M	$5/2$ $5/2$	$5/2$ $3/2$	$3/2$ $3/2$	$5/2$ $1/2$	$3/2$ $1/2$	$5/2$ $-1/2$	$3/2$ $-1/2$	$5/2$ $-3/2$	$3/2$ $-3/2$	$5/2$ $-5/2$
2	1/2			1									
2	−1/2				$\sqrt{1/5}$	$\sqrt{4/5}$							
1	1/2				$\sqrt{4/5}$	$-\sqrt{1/5}$							
1	−1/2						$\sqrt{2/5}$	$\sqrt{3/5}$					
0	1/2						$\sqrt{3/5}$	$-\sqrt{2/5}$					
0	−1/2								$\sqrt{3/5}$	$\sqrt{2/5}$			
−1	1/2								$\sqrt{2/5}$	$-\sqrt{3/5}$			
−1	−1/2										$\sqrt{4/5}$	$\sqrt{1/5}$	
−2	1/2										$\sqrt{1/5}$	$-\sqrt{4/5}$	
−2	−1/2												1

71

1 × 1

m_1	m_2	J M	2 2	2 1	1 1	2 0	1 0	0 0	2 -1	1 -1	2 -2
1	1		1								
1	0			$\sqrt{1/2}$	$\sqrt{1/2}$						
0	1			$\sqrt{1/2}$	$-\sqrt{1/2}$						
1	-1					$\sqrt{1/6}$	$\sqrt{1/2}$	$\sqrt{1/3}$			
0	0					$\sqrt{2/3}$	0	$-\sqrt{1/3}$			
-1	1					$\sqrt{1/6}$	$-\sqrt{1/2}$	$\sqrt{1/3}$			
0	-1								$\sqrt{1/2}$	$\sqrt{1/2}$	
-1	0								$\sqrt{1/2}$	$-\sqrt{1/2}$	
-1	-1										1

ISOSPIN

$3/2 \times 1$

m_1	m_2	J=5/2, M=5/2	5/2, 3/2	3/2, 3/2	5/2, 1/2	3/2, 1/2	1/2, 1/2	5/2, -1/2	3/2, -1/2	1/2, -1/2	5/2, -3/2	3/2, -3/2	5/2, -5/2
3/2	1	1											
3/2	0		$\sqrt{2/5}$	$\sqrt{3/5}$									
1/2	1		$\sqrt{3/5}$	$-\sqrt{2/5}$									
3/2	-1				$\sqrt{1/10}$	$\sqrt{2/5}$	$\sqrt{1/2}$						
1/2	0				$\sqrt{3/5}$	$\sqrt{1/15}$	$-\sqrt{1/3}$						
-1/2	1				$\sqrt{3/10}$	$-\sqrt{8/15}$	$\sqrt{1/6}$						
1/2	-1							$\sqrt{3/10}$	$\sqrt{8/15}$	$\sqrt{1/6}$			
-1/2	0							$\sqrt{3/5}$	$-\sqrt{1/15}$	$-\sqrt{1/3}$			
-3/2	1							$\sqrt{1/10}$	$-\sqrt{2/5}$	$\sqrt{1/2}$			
-1/2	-1										$\sqrt{3/5}$	$\sqrt{2/5}$	
-3/2	0										$\sqrt{2/5}$	$-\sqrt{3/5}$	
-3/2	-1												1

CHAPTER 4

Strangeness

IN 1947 THE pion was discovered, giving weight to Yukawa's theory, and the notion became current among physicists that they had now found all the "elementary particles" of which matter and energy are composed. This, however, was very far from the truth, and it was in the same year that the first of a new set of particles was found. These were called "strange particles" because of their unusual properties. Below I shall describe first these properties and then the theory that accounts for them. The chapter ends with a discussion of the parity of strange particles.

4.1 THE DISCOVERY OF STRANGE PARTICLES

The first strange particle reaction was observed in 1947 by Rochester and Butler of Manchester University, during a study of cosmic rays using cloud chambers. A typical decay observed by them is shown schematically in Figure 1. Here a solid line represents a charged particle, which leaves an ionisation track in the cloud chamber. A dashed line represents a neutral particle, which leaves no track. In Figure 1 an unknown particle decays into a proton and a negative pion. From the analysis of several such decays it was concluded that the neutral particle was produced in the lead plate placed across the chamber. By measuring the track curvature (the chamber is in a magnetic field) and ionisation densities, the energy and momentum of p and π^- can be calculated, so knowing the rest masses, the rest mass of the decaying particle may be found.

During the next few years several similar decays were observed, the most interesting of which was in 1953 by Cowan of the Pasadena group. He observed

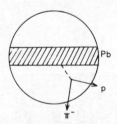

Figure 4.1 A decay of the type observed by Rochester and Butler.

STRANGENESS

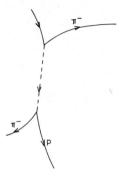

Figure 4.2

a decay shown schematically in Figure 2. This takes place in two stages. The unknown negative particle first decays into a neutral particle and a π^-, and the neutral particle then decays into p and π^-, as in Figure 1.

There was a great leap forward when accelerators were able to produce these particles in the laboratory. The first laboratory production was performed on the Brookhaven cosmotron in 1954. Protons were bombarded with negative pions accelerated to 1.5 GeV, to give

$$\pi^- + p \rightarrow \Lambda + K^0 \tag{4-1}$$

and these two neutral particles, which are now called Λ (lambda) and K^0, subsequently decayed, as shown in Figure 3. The decays

$$\Lambda \rightarrow p + \pi^- \tag{4-2}$$

and

$$K^0 \rightarrow \pi^+ + \pi^- \tag{4-3}$$

were recognised from the cosmic ray events. The Λ decay is the same one as in Figure 1, and the K^0 decay, which had also been observed, has a very similar "V"- type appearance. (For this reason, these particles were originally called V_1 and V_2, but I prefer to call them consistently by their modern names, to reduce confusion.)

Besides reaction (1) the reaction

$$\pi^- + p \rightarrow \Sigma^- + K^+ \tag{4-4}$$

was also observed, followed by the decays

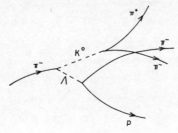

Figure 4.3

$$\Sigma^- \to n + \pi^- \tag{4-5}$$

$$K^+ \to \pi^+ + \pi^0 \tag{4-6}$$

(Σ is pronounced sigma). In due time, Σ^+ and K^- were also discovered, through their decay modes

$$\Sigma^+ \to n + \pi^+, \; \Sigma^+ \to p + \pi^0 \tag{4-7}$$

$$K^- \to \pi^- + \pi^0 \tag{4-8}$$

and the "cascade" particle with the two-stage decay of Figure 2 was named Ξ^- (ksi minus)

$$\Xi^- \to \Lambda + \pi^-$$
$$\hookrightarrow p + \pi^-. \tag{4-9}$$

The masses of all these new particles are shown in Table 1. Σ^+ and Σ^- are given the same symbol because their nearly equal masses suggests that they are members of the same isospin family — we shall return to this point. Also, K^+ and K^- have identical masses because one is the antiparticle of the other. This follows from the fact that the decay products in (8) are the antiparticles of those in (6).

Table 4.1

Particle	Mass (MeV)
Λ	1115.4
Σ^-	1197.2
Σ^+	1189.4
Ξ^-	1320.8
K^0	497.8
K^+	493.8
K^-	493.8

4.2 THE PARADOX WHICH LEAD TO THE INTRODUCTION OF THE STRANGENESS QUANTUM NUMBER

Now that we have a list of the new particles, let me make the crucial observation about their production and decay. Reaction (1) is typical, so let us consider that. At the energy of the Brookhaven experiment the total cross section for $\pi^- p$ scattering was found to be 34 ± 3 mb, while the cross section for reaction (1) was about 1 mb. In other words, (1) is certainly a strong interaction — the new particles interact strongly with pions and nucleons. On the other hand, the decays (2), (3) etc. occur very slowly, as can be seen from the tracks in Figure 3. By measuring the distance traversed by Λ and K, the half-lives for $\Lambda \to p + \pi^-$ and $K^0 \to \pi^+ + \pi^-$ were found to be 1.8×10^{-10} sec. and 0.7×10^{-10} sec. respectively. These times are typical of decays due to weak interactions. But if Λ and K^0 are produced from π and N by a strong interaction, why don't they also decay via a strong interaction? It appears that for some reason Λ and K^0 are deliberately boycotting the strong interaction in their decay. Why is this?

I should mention that this paradox was evident even before the laboratory production of strange particles, and an important key to the solution was the hypothesis of *associated production* proposed by Nambu, Nishijima and Yamaguchi, and by Oneda in 1951 and by Pais in 1952. They predicted that strange particles were produced in pairs, but decay singly. This amounts to saying that strong interactions must involve pairs of strange particles, whereas weak interactions involve only single particles. Thus a kind of selection rule operates. All of the above reactions respect this rule except (8), where the weak decay $\Xi^- \to \Lambda + \pi^-$ involves two strange particles. Even apart from this violation, however, we may reasonably object that the hypothesis is rather *ad hoc*; it looks more like a rule of thumb than an "explanation".

The situation was greatly clarified by the introduction of a new quantum number, which has come to be called *strangeness*, denoted S. Every particle has a particular (integral) value of S, and pions and nucleons have $S = 0$. The new particles, Λ, K, Σ, etc., have $S \neq 0$, and are therefore called "strange particles". The strangeness of two particles is just the sum of their strangeness; in other words, strangeness "adds", just like electric charge Q. Now the idea is that *strong interactions conserve strangeness, but weak interactions change it.* So, in reaction (1), we (arbitrarily) assign to Λ, $S = -1$, and to K^0, $S = +1$, so the net strangeness on both sides of the reaction is zero, and strangeness is conserved. The decays (2) and (3), however, clearly involve a change of strangeness, and it is for this reason that the decays of strange particles are weak. The selection rule referred to above is now a lot clearer. By assigning S to the different particles as in Table 2, we can account for the processes (1) to (9). But it is easy to see that these are not sufficient to determine S. For example, looking at (4), (5) and (6), we see that Σ^- and K^+ have equal and opposite values of S, so that the sum is zero, but we don't know whether Σ^- has $S = -1$ and K^+ has $S = +1$, or *vice versa*. In fact, may it

Table 4.2.
Provisional assignment of strangeness.

Particle	Strangeness S
N	0
π	0
Λ	-1
K^0	$+1$
Σ^-	± 1
K^+	± 1
Σ^+	± 1
K^-	± 1
Ξ^-	-2

even be possible that Σ^- has $S = -2$ and K^+ has $S = +2$ (or *vice versa*)?

We can at least answer this last question by looking at (9). Ξ^- does not decay directly into $p + 2\pi^-$ but first decays into $\Lambda + \pi^-$, and Λ then decays into $p + \pi^-$. Why is this? An obvious explanation is to assign $S = -2$ to Ξ^-, and make the rule that

Weak interactions change S by one unit only, $|\Delta S| = 1$ (4-10)

so that Ξ^- cannot decay directly into $p + 2\pi^-$. Returning to (4), (5) and (6), since K^+ and Σ^- decay in one stage to pions and nucleons, they must have $S = \pm 1$, and so must K^- and Σ^+ in (7) and (8).

4.3 ISOSPIN OF STRANGE PARTICLES AND THE GELL-MANN–NISHIJIMA RELATION

We saw in the last chapter that there is a quantum number isospin, with a vector addition rule, which is conserved in the strong interactions between pions and nucleons. Let us now extend this to assume that

Isospin is conserved in the strong interactions of strange particles   (4-11)

and, as a consequence of this, that

Isospin is a good quantum number for strange particles, which therefore exist in isospin multiplets (charge families), whose members have almost identical mass.   (4-12)

Σ and Λ particles

With (12) in mind, consider Σ^+, Λ and Σ^-. Could they be members of an $I = 1$ multiplet? Since m_Λ is some 80 MeV lower than m_{Σ^+} and m_{Σ^-}, we must answer no, and therefore (a) predict Σ^0, with $m_{\Sigma^0} \approx m_{\Sigma^\pm}$, to make up an $I = 1$ Σ triplet, and (b) assign $I = 0$ to Λ, since there are no other charged particles with a similar mass. Σ^0 has since been found, with a mass of 1192.5 MeV, and an electromagnetic decay mode

$$\Sigma^0 \to \Lambda + \gamma \tag{4-13}$$

which explains why it wasn't seen before — it decays fairly rapidly into Λ. We now have

$$\Sigma \text{ isospin } I = 1 \begin{array}{l} I_3 = +1 \ \Sigma^+ \\ I_3 = 0 \ \ \Sigma^0 \\ I_3 = -1 \ \Sigma^- \end{array} \tag{4-14}$$

and

$$\Lambda \text{ isospin } I = 0 \text{——} I_3 = 0 \ \ \Lambda \tag{4-15}$$

Let us also note that, from reactions (2), (5), (7) and (13) the baryon number B (which is conserved by *all* interactions) of Λ and Σ is $B = 1$. Λ has $S = -1$, but we still do not know (see Table 2) whether Σ^+, Σ^0 and Σ^- have $S = -1$ or $S = +1$. This is decided by the Gell-Mann–Nishijima relation, which will now be described.

We saw in the last chapter (see (3-4), (3-30) and (3-48)) that for the nucleon, pion and deuteron we have the relations

$N: (B = 1) \qquad Q = I_3 + \tfrac{1}{2}$

$\pi: (B = 0) \qquad Q = I_3$

$d: (B = 2) \qquad Q = I_3 + 1.$

These may clearly be combined, for these non-strange particles, into the one relation

$$Q = I_3 + \tfrac{1}{2} B \tag{4-16}$$

by taking baryon number into account. Now Λ has $I_3 = 0$, $B = 1$, $Q = 0$, and so since $S = -1$, we may generalise (16) to

$$Q = I_3 + \tfrac{1}{2}(B + S) \tag{4-17}$$

to account for Λ. This is called the Gell-Mann–Nishijima relation. It is satisfied by definition for N, π, d and Λ, but we shall now see that it results in satisfactory and consistent assignments for all the other particles. The quantity $(B + S)$ is sometimes called "hypercharge" and denoted Y, and (17) is then

$$Q = I_3 + Y/2. \tag{4-18}$$

The Gell-Mann–Nishijima relation predicts $S = -1$ for the Σ multiplet; for example Σ^+ has $Q = 1$, $I_3 = 1$, $B = 1$, so (17) demands $S = -1$. For ease of reference the strange particle quantum numbers and masses are displayed in Table 3.

Before leaving Σ and Λ it is worth making two observations. (a) Particle and antiparticle have opposite values of S as well as of Q and B. (b) The masses of Σ^+, Σ^0 and Σ^- are the same to within less than 1%. The hypothesis that charge independence extends to the strange particles ((11) and (12)) seems fully vindicated.

Table 4.3

Particle	B	S	I_3	Mass MeV	I
π^+	0	0	1	139.6	
π^0	0	0	0	135.0	1
π^-	0	0	-1	139.6	
K^+	0	1	½	493.8	½
K^0	0	1	-½	497.8	
$\overline{K^0}$	0	-1	½	497.8	½
K^-	0	-1	-½	493.8	
p	1	0	½	938.3	½
n	1	0	-½	939.6	
Λ	1	-1	0	1115.4	0
Σ^+	1	-1	1	1189.4	
Σ^0	1	-1	0	1192.3	1
Σ^-	1	-1	-1	1197.2	
Ξ^0	1	-2	½	1314.3	½
Ξ^-	1	-2	-½	1320.8	

K mesons

Returning again to reaction (1), note that since it is a strong interaction, then by (11) it conserves isospin. Since $\pi^- + p$ has $I = \frac{3}{2}$ or $\frac{1}{2}$, then so has $\Lambda + K^0$. Since Λ has $I = 0$, this means that K has $I = \frac{3}{2}$ or $\frac{1}{2}$, and that K^0 must be a member of a family of either 4 or 2 particles. Since K^0 has $S = 1$ by our original assignment, (17) gives $I_3 = -\frac{1}{2}$, so the complete multiplet must be either

$$I = \tfrac{3}{2} \qquad K^{++} \quad K^+ \quad K^0 \quad K^-$$
$$I_3: \qquad \tfrac{3}{2} \quad \tfrac{1}{2} \quad -\tfrac{1}{2} \quad -\tfrac{3}{2}$$

or

$$I = \tfrac{1}{2} \qquad K^+ \quad K^0$$
$$I_3: \quad \tfrac{1}{2} \quad -\tfrac{1}{2}$$

K^+ has been found, as in reaction (4), and since Σ^- has $S = -1$, then K^+ has $S = 1$, the same as K^0, as required by either of the above schemes. However, a K^{++} with mass about 495 MeV has never been found, so we deduce that K^+ and K^0 are the two members of a multiplet with $I = \frac{1}{2}$ and $S = 1$

$$K \text{ isospin } I = \tfrac{1}{2} \begin{cases} I_3 = \tfrac{1}{2} & K^+ \\ I_3 = -\tfrac{1}{2} & K^0 \end{cases} \qquad (4\text{-}19)$$

Their masses are 493.8 and 497.8 MeV, sufficiently close for the difference to be credibly due to electromagnetic forces only.

The particles (K^+, K^0) will have antiparticles $(K^-, \overline{K^0})$ with $S = -1$, so K^0 is *not* its own antiparticle, as π^0 is. This has very remarkable consequences, which are described in chapters 8 and 9. Also K^+ and K^- must have identical masses, as is indeed found experimentally. Lastly, K^- has $S = -1$, so it is not produced together with Λ and Σ in reactions like (1) and (4). This would be forbidden by strangeness conservation in strong interactions. This explains the originally observed preponderance of K^+ over K^- mesons. (Note that if K had $I = \frac{3}{2}$, then K^- would have $S = 1$, and this preponderance would be unexplained.) In summary we have

$$\overline{K} \text{ isospin } I = \tfrac{1}{2} \begin{cases} I_3 = \tfrac{1}{2} & \overline{K}^0 \\ I_3 = -\tfrac{1}{2} & K^- \end{cases} \qquad (4\text{-}20)$$

The cascade particle

Having previously agreed that Ξ^- has $S = -2$, and knowing that it has $Q = -1$ and

82   ELEMENTARY PARTICLES AND SYMMETRIES

$B = 1$ (since B is exactly conserved in (9)), we invoke (17) to give $I_3 = -\tfrac{1}{2}$. The minimum I consistent with this is $I = \tfrac{1}{2}$, and the particle with $I_3 = +\tfrac{1}{2}$ would be a Ξ^0. This particle has since been found, and the production reactions on the Berkeley bevatron

$$K^- + p \to \Xi^- + K^+$$

$$K^- + p \to \Xi^0 + K^0$$

are consistent with these quantum numbers. In addition, a Ξ^{--} and Ξ^+ have never been found, so by exactly the same reasoning that we used for the K mesons, the assignments $I = \tfrac{3}{2}$ or higher are ruled out. We have

$$\Xi \text{ isospin } I = \tfrac{1}{2} \begin{cases} I_3 = \tfrac{1}{2} \; \Xi^0 \\ I_3 = -\tfrac{1}{2} \; \Xi^- \end{cases} \tag{4-21}$$

The masses of Ξ^- and Ξ^0 are the same to within 1%, thus again providing strong support for the hypothesis of isospin conservation in the strong interactions of strange particles.

These results are summarised in Table 3. The neat way in which the strange particles, which at first caused so much puzzlement, fall into families of definite isospin, is a beautiful and powerful vindication of the idea of isospin conservation and its consequences.

4.4 PARITY OF THE K MESON

The parity of K^- is determined from the occurrence of the reaction

$$K^- + He^4 \to {}_\Lambda H^4 + \pi^0$$

where ${}_\Lambda H^4$ is a so-called "hyperfragment" of H^4, with one of the neutrons replaced by a Λ. That is, ${}_\Lambda H^4 = (pnn\Lambda)$. This reaction is then equivalent to $K^- + p \to \Lambda + \pi^0$. K^-, He^4 and π^0 are all known to have spin zero. It can be argued that the spin of ${}_\Lambda H^4$ is also zero. In that case, conservation of total angular momentum implies conservation of orbital angular momentum, since no spin is involved. So $(l)_{\text{lhs}} = (l)_{\text{rhs}} = l$, and parity conservation then implies that

$$\epsilon(K^-)(-1)^l = \epsilon(\pi^0)(-1)^l$$

i.e.

$$\epsilon(K^-) = \epsilon(\pi^0) = -1,$$

where we have used the convention (2-28), $\epsilon(p) = \epsilon(n) = \epsilon(\Lambda) = 1$, to deduce the fact that He^4 and $_\Lambda H^4$ have positive parity.

It is obviously crucial to this argument that $_\Lambda H^4$ has spin zero. The situation is actually a little unclear, but comparison of theory and experiment certainly favours spin 0 at present.

We deduce that K^- has negative parity, and, since kaons are bosons, that K^+ also does; and from isospin, that K^0 and $\bar{K}^0$ do too. In other words, K and $\bar{K}$, like π, are 0^- (pseudoscalar) particles. (Note that the way of stating this result without reference to the convention (2-28) is to say that *the K-N-Λ relative parity is odd.*)

4.5 PARITY OF Σ

Evidence for the parity of Σ comes from study of their production and decay

$$K^- + p \to \Sigma^{\pm,0} + \pi^{\mp,0}, \quad \Sigma \to N + \pi$$

and the asymmetry of the Σ decay relative to the production plane is the quantity to be measured. The analysis is rather tedious, so I shall simply state that the experimental evidence, though subject to ambiguities, suggests† that the K-N-Σ relative parity is odd, or, using convention (2-28), that Σ have positive parity.

It is also worth mentioning that an independent test exists for the $\Sigma^0 - \Lambda$ relative parity, by measuring the rate for

$$\Sigma^0 \to \Lambda + e^+ + e^-.$$

This an electromagnetic process, the "internal conversion" analogue of $\Sigma^0 \to \Lambda + \gamma$. Again, the evidence‡ suggests an even $\Sigma^0 - \Lambda$ relative parity.

4.6 PARITY OF Ξ

The parity of Ξ^- may be determined from a study of the reaction

$$\Xi^- + p \to 2\Lambda.$$

If the initial and final orbital angular momenta are l_i and l_f, we have

$$\epsilon(\Xi^-)\epsilon(p)(-1)^{l_i} = (\epsilon(\Lambda))^2 (-1)^{l_f}$$

---

† For further details, see K. Nishijima, *Fundamental Particles,* Benjamin, 1963, p. 303.
‡ For further details, see S. Gasiorowicz, *Elementary Particle Physics,* Wiley, 1967, p.220.

or

$$\epsilon(\Xi^-) = (-1)^{l_i+l_f} \epsilon(p).$$

The interesting point is that the $\Xi^- - p$ relative parity may be determined *independently* of the Λ parity. This is because Ξ^- and p differ in strangeness by an even number. By measuring l_i and l_f the relative $\Xi^- - p$ parity may be determined. The experiment has not yet been performed.

Because of isospin symmetry, it is believed that Ξ^0 and Ξ^- have the same parity (relative to p and n).

FURTHER READING

Gell-Mann, M. and A. H. Rosenfeld, "Hyperons and Heavy Mesons (systematics and decay)", *Annual Reviews of Nuclear Science,* **7**, 407 (1957).

Kemmer, N., J. C. Polkinghorne and D. L. Pursey, "Invariance in Elementary Particle Physics", *Reports on Progress in Physics,* **22**, 368 (1959).

Nishijima, K., *Fundamental Particles,* Benjamin, 1963, Chapter 6.

CHAPTER 5

Isospin and Strangeness Selection Rules in Weak and Electromagnetic Interactions

5.1 INTRODUCTION: LEPTONIC AND NONLEPTONIC WEAK INTERACTIONS

IN THE PREVIOUS chapters we have considered the quantum numbers isospin and strangeness, and have seen that they are both conserved by strong interactions. We saw, however, that electromagnetic interactions do not conserve I, and that weak interactions (or at least some of them) do not conserve S. A question which then arises is, do these interactions change I and S in a *particular* or systematic way? A part of this question is already answered for us, since we saw that to account for the two-stage decay of the cascade particle

$$\Xi^- \to \Lambda + \pi^-$$
$$\downarrow$$
$$p + \pi^-$$

we had to introduce the rule

$|\Delta S| = 1$ only

for these weak interactions. This is a *selection rule* for strangeness which is obeyed by the strangeness violating nonleptonic (see below) weak interactions. In this chapter we want to find what selection rules there are which describe how all weak and electromagnetic interactions behave with respect to isospin and strangeness.

Before doing this, however, let me remark that the idea of selection rules is by no means a new one, but should be familiar from atomic physics. The most common *radiative transitions* in atoms are of the electric dipole type, and the selection rules which govern these transitions in a one-electron atom, ignoring spin, are

$\Delta l = \pm 1, \quad \Delta m = \pm 1, 0, \quad \Delta n$ unrestricted.

The selection rules that we want are analogous, except that we are not concerned

86 ELEMENTARY PARTICLES AND SYMMETRIES

with space-time quantum numbers like angular momentum, but with isospin and strangeness.

There are two types of weak interactions, those which involve neutrinos, and those which don't. Those which do, also involve e^- or μ^- (or e^+ or μ^+), and a typical example is the beta decay of the neutron

$$n \to p + e^- + \bar{\nu}_e.$$

Because leptons are involved, these are called *semi-leptonic* (or sometimes simply *leptonic*) weak interactions. Weak interactions which do not involve neutrinos are called *nonleptonic*, since they involve no leptons at all. We saw many examples of these in the last chapter, for example

$$\Lambda \to p + \pi^-.$$

Although semi-leptonic and nonleptonic weak interactions are of roughly the same strength, they may be rather different in origin, so in what follows I shall treat them separately.

Finally, there are weak interactions which involve leptons *only*. These are called *purely leptonic* weak interactions. The only purely leptonic decays are

$$\mu^- \to e^- + \bar{\nu}_e + \nu_\mu$$

and

$$\mu^+ \to e^+ + \nu_e + \bar{\nu}_\mu$$

but the elastic scattering

$$e^- + \bar{\nu}_e \to e^- + \bar{\nu}_e$$

(which has not yet been performed experimentally), is also a purely leptonic weak interaction. We shall not be concerned with this sub-class in this chapter, since isospin and strangeness do not exist for leptons, and we are interested in precisely the selection rules for these quantum numbers.

5.2 STRANGENESS CONSERVING SEMILEPTONIC WEAK INTERACTIONS

The first example of a strangeness conserving weak decay is

$$n \to p + e^- + \bar{\nu}_e. \tag{5-1}$$

ISOSPIN AND STRANGENESS SELECTION RULES

$I \quad ½ \to ½$

$I_3 \quad -½ \to ½$

$S \quad 0 \to 0$

It is clear that this decay involves no change of strangeness; we write $\Delta S = 0$. Now

$$\Delta \mathbf{I} = \mathbf{I}_f - \mathbf{I}_i \tag{5-2}$$

$$\Delta I_3 = (I_3)_f - (I_3)_i, \tag{5-3}$$

where f and i stand for final and initial states. (2) involves using the vector addition rule, by which the vector sum of $\mathbf{I}_f$ and $\mathbf{I}_i$ takes on all values from $I_f + I_i$ to $|I_f - I_i|$, in steps of 1. This is exactly the same as the vector sum of $\mathbf{I}_f$ and $-\mathbf{I}_i$; in other words, the *vector difference of two vectors is the same as their vector sum*. So we may rewrite (2)

$$\Delta \mathbf{I} = \mathbf{I}_f + \mathbf{I}_i. \tag{5-4}$$

We also have

$$|\Delta I_3| \leq \Delta I, \tag{5-5}$$

where $\Delta I = |\Delta \mathbf{I}|$. This follows since the component of a vector along any axis can never exceed the length of the vector.

Returning to neutron decay, we have

$$\Delta I_3 = ½ - (-½) = 1. \tag{5-6}$$

According to (4), on the other hand, ΔI is the vector sum of ½ and ½ which is[†]

$$\Delta I = 0 \text{ or } 1, \tag{5-7}$$

but in view of (5) and (6), we have

$$\Delta I = 1 \text{ only}. \tag{5-8}$$

[†] Perhaps the reader may find the intuitive derivation of (7) more helpful. From (2), which we write in the form $\mathbf{I}_i + \Delta \mathbf{I} = \mathbf{I}_f$, we see that $\Delta \mathbf{I}$ is the vector we have to add to $\mathbf{I}_i$ to get $\mathbf{I}_f$, using the vector addition rule. In the case of neutron beta decay (1), then, $\Delta \mathbf{I}$ is any vector which, added vectorially to $I = 1/2$ will give $I = 1/2$. Both 0 and 1 will do this, since $1/2 + 0 = 1/2$, and $1/2 + 1 = 1/2$ or $3/2$. This gives the same result as (8), by a longer method.

Writing (6) and (8) together,

$$\Delta I = 1, \quad \Delta I_3 = 1. \tag{5-9}$$

Now consider the decay

$$\pi^+ \to \mu^+ + \nu_\mu. \tag{5-10}$$

I 1

I_3 1

This is the most common decay mode of the charged pion. The right hand side contains only particles for which isospin and strangeness are undefined, so we have by inspection

$$\Delta I = 1, \quad \Delta I_3 = -1, \tag{5-11}$$

and of course $\Delta S = 0$.

Now consider the rather rare decay mode of the charged pion

$$\pi^+ \to \pi^0 + e^+ + \nu_e$$

I 1   1

I_3 1   0

and $\Delta S = 0$. By (3) we have

$$\Delta I_3 = -1, \tag{5-13}$$

and ΔI is the vector sum of 1 and 1, that is

$$\Delta I = 0, 1, 2. \tag{5-14}$$

Comparison with (13) then gives

$$\Delta I = 1 \text{ or } 2, \quad \Delta I_3 = -1. \tag{5-15}$$

What this means is that there are two possible ways in which this reaction can take place. There is an amplitude that it happens through an interaction which changes I by 1, and there is another amplitude that it takes place through an interaction which changes I by 2. Unfortunately, it is not possible to disentangle

ISOSPIN AND STRANGENESS SELECTION RULES 89

these amplitudes experimentally. In view of (9) and (10), however, it is very attractive to make the hypothesis

Strangeness conserving semileptonic weak interactions obey the selection rules
$$\Delta S = 0, \quad \Delta I = 1, \quad \Delta I_3 = \pm 1. \tag{5-16}$$

Wherever testable, this hypothesis has been found to work.

5.3 STRANGENESS CHANGING SEMILEPTONIC WEAK INTERACTIONS

We proceed exactly as before, considering the decays of strange particles. First, the decay

$$\Lambda \to p + e^- + \bar{\nu}_e \tag{5-17}$$

I	0	½
I_3	0	½
S	−1	0
Q	0	1

clearly has the selection rules

$$\Delta S = \Delta Q = 1, \quad \Delta I = \text{½}. \tag{5-18}$$

$\Delta Q = 1$ does not mean, of course, that charge is not conserved; only that the charge on the hadrons is not conserved.
Next, the decay

$$\Sigma^- \to n + e^- + \bar{\nu}_e \tag{5-19}$$

I	1	½
I_3	−1	−½
S	−1	0
Q	−1	0

obeys

$$\Delta S = \Delta Q = 1, \quad \Delta I_3 = \text{½}, \tag{5-20}$$

and, by the vector addition rule

$$\Delta I = 1/2, 3/2. \tag{5-21}$$

The decay (19) is very rare, and accounts for less than 0.1% of all Σ^- decays. However, the corresponding decay of Σ^+ has never been seen (strictly, it accounts for less than 2×10^{-5} of all Σ^+ decays)

$$\Sigma^+ \to n + e^+ + \bar{\nu}_e \tag{5-22}$$

I	1	1/2
I_3	1	$-1/2$
S	-1	0
Q	1	0

By using the techniques as above, this decay would have

$$\Delta S = -\Delta Q = 1, \quad \Delta I = 3/2 \text{ (forbidden)} \tag{5-23}$$

where $\Delta I = 3/2$ because $\Delta I_3 = -3/2$.

We interpret the fact that $\Sigma^+ \not\to n + e^+ + \nu_e$ by saying that decays with $\Delta I = 3/2$ are *forbidden*. It is then reasonable to assume that in the decay (19) only the amplitude with $\Delta I = 1/2$ makes any contribution to the reaction. Collecting the information from the leptonic decays of strange baryons, we have the selection rule

$$\Delta S = \Delta Q = 1, \quad \Delta I = 1/2. \tag{5-24}$$

Let us now consider the decays of the strange mesons K. In close parallel with the pion decays (10) and (12) there are

$$K^+ \to \mu^+ + \nu_\mu; \quad K^- \to \mu^- + \bar{\nu}_\mu \tag{5-25a, b}$$

I	1/2	1/2
I_3	1/2	$-1/2$
S	1	-1
Q	1	-1

ISOSPIN AND STRANGENESS SELECTION RULES

and

$$K^+ \to \pi^0 + \mu^+ + \nu_\mu \quad K^- \to \pi^0 + \mu^- + \bar{\nu}_\mu \quad (5\text{-}26a, b)$$

I	½	1	½	1
I_3	½	0	–½	0
S	1	0	–1	0
Q	1	0	–1	0

By inspection, (25) obey

$$\Delta S = \Delta Q = -1, \quad \Delta I = \tfrac{1}{2}; \quad \Delta S = \Delta Q = 1, \quad \Delta I = \tfrac{1}{2}. \quad (5\text{-}27a, b)$$

By reasoning which is now familiar, the decays (26) both have $\Delta I = \tfrac{1}{2}$ or $\tfrac{3}{2}$. It can be shown, however, that if $\Delta I = \tfrac{1}{2}$ only, the following relation holds between the rates of leptonic decays (K_2^0 is a linear superposition of K^0 and $\overline{K^0}$ to be discussed in Chapter 9)

$$\Gamma(K_2^0 \to \pi^- + \mu^+ + \nu_\mu) + \Gamma(K_2^0 \to \pi^+ + \mu^- + \bar{\nu}_\mu) = 2\Gamma(K^+ \to \pi^0 + \mu^+ + \nu_\mu). \quad (5\text{-}28)$$

Relation (28) has been verified experimentally, within the errors. In other words, the amplitude with $\Delta I = \tfrac{3}{2}$ does not contribute, and we have for the reactions (26)

$$\Delta S = \Delta Q = -1, \quad \Delta I = \tfrac{1}{2}; \quad \Delta S = \Delta Q = 1, \quad \Delta I = \tfrac{1}{2}. \quad (5\text{-}29a, b)$$

These are the same as (27).

Finally, the only leptonic decay of Ξ that has been seen with any certainty is

$$\Xi^- \to \Lambda + e^- + \bar{\nu}_e. \quad (5\text{-}30)$$

In particular, the decay $\Xi^- \to n + e^- + \bar{\nu}_e$ has not been seen[†], so we may reasonably expect that for the leptonic decays of Ξ,

$$\Delta S = \Delta Q = 1, \quad \Delta I = \tfrac{1}{2} \quad (5\text{-}31)$$

---

† Experimentally, $\dfrac{\text{rate}(\Xi^- \to ne^-\bar{\nu}_e)}{\text{rate}(\Xi^- \to \Lambda e^-\bar{\nu}_e)} < \tfrac{1}{2}$

with $\Delta S = 2$ forbidden, exactly as for *nonleptonic* Ξ decay, as mentioned in the introduction to this chapter.

In conclusion, we may state with a reasonable degree of confidence that

Strangeness changing leptonic weak interactions obey the selection rules
$$\Delta S = \Delta Q = \pm 1, \quad \Delta I = \tfrac{1}{2}. \tag{5-32}$$

EXERCISE   Using the Gell-Mann–Nishijima relation, prove that for leptonic decays

$$\Delta S = -\Delta Q \text{ implies that } \Delta I = \tfrac{3}{2}. \tag{5-33}$$

5.4   NONLEPTONIC WEAK INTERACTIONS

In the last chapter we saw examples of many leptonic weak interactions, since they played a key role in the introduction of strangeness. For convenience let me list the decays here:

$$\Lambda \to p + \pi^- \tag{4-34}$$

$$\Sigma^+ \to n + \pi^+ \tag{5-35}$$

$$\Sigma^+ \to p + \pi^0 \tag{5-36}$$

$$\Sigma^- \to n + \pi^- \tag{5-37}$$

$$\Xi^- \to \Lambda + \pi^- \tag{5-38}$$

$$K^0 \to \pi^+ + \pi^- \tag{5-39}$$

$$K^+ \to \pi^+ + \pi^0. \tag{5-40}$$

It is easy to check that all of these decays obey[†]

$$\Delta S = -2\Delta I_3 = \pm 1. \tag{5-41}$$

The only question of further issue is whether they obey $\Delta I = \tfrac{1}{2}$ or $\Delta I = \tfrac{3}{2}$, or both. This is decided by comparing the decay rates into different final states.

---

† The degree of certainty that $\Delta S \neq 2$ in Ξ decays is given by the experimental result

$$\frac{\text{rate } (\Xi^- \to n + \pi^-)}{\text{rate } (\Xi^- \to \Lambda + \pi^-)} < 10^{-3}.$$

ISOSPIN AND STRANGENESS SELECTION RULES

For example, consider $\Lambda \to p + \pi^-$. Λ has $I = 0$, so if the decay has $\Delta I = \frac{1}{2}$, then $p + \pi^-$ is in an $I = \frac{1}{2}$ state. Exactly the same reasoning applies to the decay $\Lambda \to n + \pi^0$. But from appendix 3A, equation (3A-25) (page 68), we see that the combination

$$\sqrt{\tfrac{1}{3}}(\pi^0 n) - \sqrt{\tfrac{2}{3}}(\pi^- p)$$

is a pure $I = \frac{1}{2}$ state. In other words, the amplitude for $(\pi^- p)$ to have $I = \frac{1}{2}$ is $(-\sqrt{2})$ times the amplitude for $(\pi^0 n)$ to have $I = \frac{1}{2}$. Or, since probability $\propto |\text{amplitude}|^2$, the probability that $(\pi^- p)$ has $I = \frac{1}{2}$ is twice the probability for $(\pi^0 n)$. So if the Λ decays obey $\Delta I = \frac{1}{2}$, then

$$\frac{\text{rate}(\Lambda \to \pi^- p)}{\text{rate}(\Lambda \to \pi^0 n)} = 2$$

or

$$R = \frac{\text{rate}(\Lambda \to \pi^- p)}{\text{rate}(\Lambda \to \pi^- p) + \text{rate}(\Lambda \to \pi^0 n)} = \tfrac{2}{3}$$

The current experimental value for R is 0.653 ± 0.013, and constitutes strong evidence that $\Delta I = \frac{1}{2}$ in Λ non leptonic decay.

A similar test for the $\Delta I = \frac{1}{2}$ selection rule exists for the Σ decays (35) to (37). This again involves measuring their relative rates, and again no contradiction is found with $\Delta I = \frac{1}{2}$. The same is also true for the Ξ decays; the $\Delta I = \frac{1}{2}$ rule prediction for the relative rates of $\Xi^0 \to \Lambda + \pi^0$ and $\Xi^- \to \Lambda + \pi^-$ is in quite good agreement with experiment.

Let us now pass to the kaon decays (39) and (40). Experimentally $K^+ \to \pi^+ + \pi^0$ proceeds about 700 times faster than $K^0 \to \pi^+ + \pi^-$, and this turns out to be a consequence of the $\Delta I = \frac{1}{2}$ rule. (In fact, it was to explain this fact that the $\Delta I = \frac{1}{2}$ rule was first postulated, by Gell-Mann and Pais.) To see this, consider the general decay $K \to 2\pi$. K has $I = \frac{1}{2}$, so if the interaction causing the decay changes I by $\frac{1}{2}$, then the 2π state must have $I = 0$ or 1. On the other hand, a 2π state will obey the generalised Pauli principle (see Chapter 3), according to which the pion wave function must be symmetric under the simultaneous interchange of space, spin and isospin coordinates. The space symmetry is $(-1)^l$ where l is the relative angular momentum of the pions. Since the spins of K and π are both zero (see sections 2.4, 2.5 and 8.1), $l = 0$, so the wave function is symmetric in space coordinates. For the same reason, it is also symmetric in spin coordinates, since no spin is involved. It must therefore be symmetric in isospin. Now two pions have $I = 0$, 1 or 2, and of these $I = 0, 2$ are symmetric and $I = 1$ antisymmetric in isospin, as remarked after equation (3A-28) in Appendix 3A. The conclusion is that the allowed isospin of the final 2π state is

$I = 0, 1 \quad \Delta I = \frac{1}{2}$ rule

$I = 0, 2 \quad$ generalised Pauli principle.

If both hold, then $I = 0$ only. This straightway forbids the decay $K^+ \to \pi^+ + \pi^0$ since the final state has $I_3 = 1$, i.e. $I \geqslant 1$. It is now explained why this decay appears so depressed compared with the allowed decay $K^0 \to \pi^+\pi^-$.

Thus a $\Delta I = \frac{1}{2}$ rule holds for both baryon and meson nonleptonic decays, and

Nonleptonic weak interactions obey the selection rule
$|\Delta S| = 1, \quad \Delta I = \frac{1}{2}.$ (5-42)

EXERCISE Using the Gell-Mann–Nishijima relation, verify that for nonleptonic weak interactions $\Delta I = \frac{1}{2}$ implies that $|\Delta S| = 1$, but not vice versa.

5.5 ELECTROMAGNETIC INTERACTIONS

Reactions involving the photon γ are electromagnetic. First consider the decay

$$\pi^0 \to 2\gamma \tag{5-43}$$

$I \quad 1$

$I_3 \quad 0$

$S \quad 0$

which is by far the most common decay mode of π^0. Since γ has neither isospin nor strangeness, the decay clearly has

$\Delta I = 1, \quad \Delta I_3 = 0, \quad \Delta S = 0$ (5-44)

We have met no electromagnetic decays of the K mesons, so let us now turn to the baryons. As observed in (4-13), Σ^0 decays electromagnetically

$$\Sigma^0 \to \Lambda + \gamma \tag{5-45}$$

$I \quad 1 \quad 0$

$I_3 \quad 0 \quad 0$

ISOSPIN AND STRANGENESS SELECTION RULES   95

S -1 -1

and this reaction clearly obeys the same selection rules (44) as $\pi^0 \to 2\gamma$ does. Now consider the process

$$\gamma + d \to p + n \qquad (5\text{-}46)$$

I	0	½	½
I_3	0	½	−½
S	0	0	0

which, for obvious reasons, is called deuteron photodisintegration. In Chapter 3 we saw that d has $I = 0$. Since $(p + n)$ may have either $I = 0$ or $I = 1$, (46) will have in general

$$\Delta I = 0, 1, \quad \Delta I_3 = 0, \quad \Delta S = 0. \qquad (5\text{-}47)$$

In the two previous examples we saw that $\Delta I = 1$. This example raises the question, do electromagnetic interactions ever obey $\Delta I = 0$? The answer is they do, and a direct illustration is the decay

$$\eta \to 2\gamma. \qquad (5\text{-}48)$$

I 0

S 0

The η (eta) meson will be discussed in section 7.9, but all that needs to be known here is that it is a spin zero isospin zero uncharged meson, one of whose decay modes is (48). This clearly has

$$\Delta I = 0, \quad \Delta I_3 = 0, \quad \Delta S = 0. \qquad (5\text{-}49)$$

The selection rules (47) are believed to be universally valid for the electromagnetic interactions. They are interesting selection rules. In the first place, electromagnetic interactions conserve strangeness, so they are only effective between particles (or sets of particles) with the same strangeness. Secondly, they conserve I_3, this is related to the fact that they conserve charge − the charge Q on the hadrons, that is. But perhaps the most interesting thing about electromagnetic interactions is that, according to (47) they contain two "parts", one with $\Delta I = 0$ and the other with $\Delta I = 1$. The $\Delta I = 0$ part is responsible for $\eta \to 2\gamma$ and the $\Delta I = 1$ part for $\pi^0 \to 2\gamma$. The $\Delta I = 0$ part obeys exactly the same selection rules as strong interactions; everything is conserved, $\Delta S = \Delta I = \Delta I_3 = 0$. What then is the difference between this part of the electromagnetic interaction and the strong interaction? First there is a difference in strength; the electromagnetic interaction is about 100 times weaker than the strong interaction. But there is also a point of principle about the decay $\eta \to 2\gamma$. We say that $\Delta I = 0$ because η is a hadron with $I = 0$, and the photon is a particle *for which isospin is undefined*. Mathematically this amounts to calling $I = 0$ for the photon, but this is *not* to say that the photon is a hadron with $I = 0$. It is no such thing. Strong interactions account only for transitions between one set of hadrons and another set with the same isospin and strangeness. $\eta \to 2\gamma$ however is a transition between a hadron with $I = S = 0$ and a non-hadronic state — a very different thing.

EXERCISE Using the Gell-Mann-Nishijima relation, prove that for the electromagnetic interactions of hadrons, $\Delta S = 0$ implies $\Delta I_3 = 0$ and vice versa, but that no selection rule for ΔI may be derived from $\Delta S = 0$.

FURTHER READING

See Further Reading list for Chapter 4 (p.84)

CHAPTER 6
Electromagnetic Structure of Nucleons

WE HAVE so far been concerned with the interactions between elementary particles, but not with the internal structure of the particles themselves. In this chapter we study the structure of the proton and neutron as revealed by scattering electrons from them, both elastically and inelastically. Since their interaction with electrons is electromagnetic, the information we gain will be about the electromagnetic structure, for example, the charge distribution, of the nucleon. Besides the obvious intrinsic interest of this subject, it turns out that there are some connections between the electromagnetic structure of nucleons, and some strong interaction phenomena. Many of these matters are not yet understood and the exploration of this field is one of the most exciting aspects of present day experimental and theoretical particle physics.

6.1 THE DIFFERENCE BETWEEN ELECTRONS AND NUCLEONS

Electrons are believed to be point charges, so that when they interact with photons, they do so at a particular point in space and time. Protons, on the other hand, are surrounded by pion clouds, so that a proton may spend some of its life as a "bare" proton, but it may also sometimes exist as a neutron surrounded by a cloud of π^+, according to the transitions

$$p \leftrightarrow n + \pi^+ \qquad \textcircled{p} \leftrightarrow + \; {}^{+}_{+}\textcircled{n}{}^{+}_{+} \; +$$

Similarly, a neutron will, for some of its life exist as a proton surrounded by a cloud of π^-

$$n \leftrightarrow p + \pi^- \qquad \textcircled{n} \leftrightarrow - \; {}^{-}_{-}\textcircled{p}{}^{-}_{-} \; -$$

These pion clouds obviously give rise to a structure of the nucleons, and since the pions are charged, will result in a charge and magnetic moment distribution. In other words, there will be a "size" associated with the charge and magnetic moment, of the rough order of the pion Compton wavelength, $\sim 10^{-13}$ cm.

This difference between electrons and nucleons may be illustrated by considering electron-electron and electron-nucleon scattering. Both interactions are electromagnetic, and are described (at least to first order) by the exchange of one virtual photon. The photon will "join" onto the electron at a point, as mentioned above, so the diagram for this is as in Figure 1, which represents e-e scattering. In e-p scattering, the exchanged photon may not interact with the proton at a point, and to indicate this a "bubble" is usually drawn, as in Figure 2. The bubble is due to the pion cloud and is explained diagrammatically in Figure 3. In the first place, there is a chance that the proton may be a "bare" proton, and in this case it will behave like an electron, and interact with γ at a point. On the other hand, corresponding to $p \to n + \pi^+$, the photon γ may interact with the charged pion cloud, as shown in the second diagram on the right hand side of Figure 3. The virtual π^+, however, may itself dissociate (virtually) into say $\Sigma^+\overline{\Lambda}$, and γ may interact with Σ^+; and so on — it is easy to write down yet more complicated diagrams. The bubble in Figure 2 represents the sum of all these possibilities, which describe the non-pointlike nature of the proton.

The first consequence of the existence of pion clouds is that nucleons have anomalous magnetic moments. Dirac showed that point particles with spin ½, like the electron, have magnetic moments given by

$$\mu = \frac{Q\hbar}{2mc} \qquad (6\text{-}1)$$

where Q is the charge and m the mass. If the proton and neutron were point particles, then, we should have

$$\mu_p = \mu_N, \quad \mu_n = 0. \qquad (6\text{-}2)$$

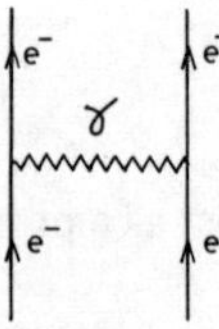

Figure 6.1 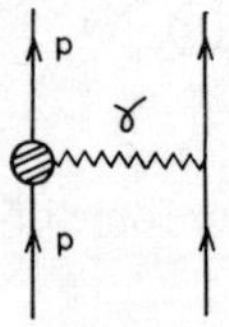   Figure 6.2

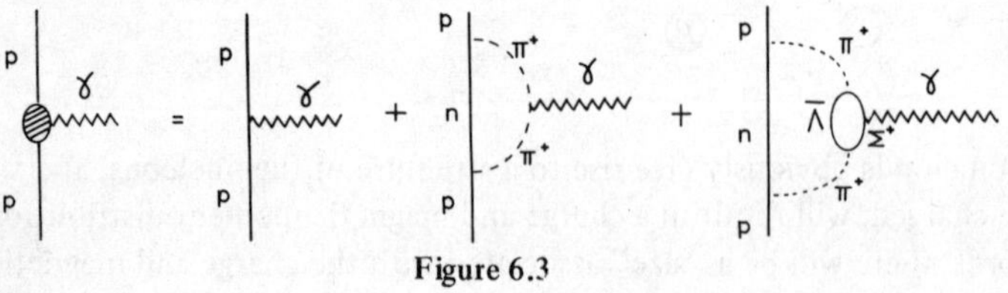

Figure 6.3

ELECTROMAGNETIC STRUCTURE OF NUCLEONS

As it is, however,

$$\mu_p = 2.78\,\mu_N, \quad \mu_n = -1.93\,\mu_N \tag{6-3}$$

where $\mu_N = e\hbar/2m_N c$ is the nuclear magneton. The discrepancy between expressions (2) and (3), called the anomalous magnetic moment, is due to the pion cloud — obviously, a circulating cloud of charged particles will give an extra contribution to the magnetic moment.

Before proceeding with the investigation of nucleons, it should be mentioned that electrons are also surrounded by clouds, in this case of virtual *photons*. These may further virtually dissociate into e^+e^- pairs which recombine into photons, $\gamma \to e^+e^- \to \gamma$, just as for pions we had $\pi^+ \to \Sigma^+\bar{\Lambda} \to \pi^+$. Because of this we expect that the electron magnetic moment will not be exactly equal to the Dirac value. However, the interaction of photons with electrons is much weaker than the interaction of pions with nucleons, so the anomalous magnetic moment should be correspondingly small. In addition, the interaction between γ and e is described by quantum electrodynamics, a completely understood theory. Writing (1) as

$$\mu = g_s \frac{e\hbar}{2mc} s$$

where s is a spin operator with eigenvalues $+\tfrac{1}{2}$ and $-\tfrac{1}{2}$, we then see that a point Dirac particle should have $g_s = 2$. Quantum electrodynamical calculations give the following perturbation series for the change in g due to photon clouds

$$\left(\frac{g-2}{2}\right)_{\text{theory}} = \frac{a}{2\pi} - 0.32848\,\frac{a^2}{\pi^2} + 0.19\,\frac{a^3}{\pi^3}$$

$$= (115964.1 \pm 0.3) \times 10^{-8} \tag{6-4}$$

where a, the fine structure constant, is $a = e^2/\hbar c = 1/137$. The experimental value is

$$\left(\frac{g-2}{2}\right)_{\text{exp}} = (115964.4 \pm 0.7) \times 10^{-8}$$

indicating a triumph both for the accuracy of the experiment and for the validity of quantum electrodynamics.

A similar calculation is impossible for the proton and neutron, however, because a perturbation series in the strong coupling constant $g_{\pi NN}$, analogous to

(4), *diverges*, since $g_{\pi NN} \simeq 14$, whereas $\alpha = 1/137$. This would give nonsense—the nucleon anomalous magnetic moments are large, but they are not infinite! Unfortunately, however, there is so far no quantitative theory of strong interactions enabling the anomalous magnetic moments to be calculated correctly.

6.2 FORM FACTORS AS A DESCRIPTION OF STRUCTURE

Let the kinematics for *e-p* scattering be as in Figure 4, where p_1, p_2, p_3, p_4 and q are 4-momenta. Energy-momentum conservation gives

$$p_1 + p_3 = p_2 + p_4 \qquad (6\text{-}5)$$

$$q = p_3 - p_4 = p_2 - p_1. \qquad (6\text{-}6)$$

From the relativistic relation between energy, momentum and mass, we have, in units where $c = 1$,

$$p_1^2 = \mathbf{p}_1 \cdot \mathbf{p}_1 - E_1^2 = -m_p^2 \qquad (6\text{-}7a)$$

$$p_2^2 = \mathbf{p}_2 \cdot \mathbf{p}_2 - E_2^2 = -m_p^2 \qquad (6\text{-}7b)$$

$$p_3^2 = \mathbf{p}_3 \cdot \mathbf{p}_3 - E_3^2 = -m_e^2 \qquad (6\text{-}7c)$$

$$p_4^2 = \mathbf{p}_4 \cdot \mathbf{p}_4 - E_4^2 = -m_e^2 \qquad (6\text{-}7d)$$

On the other hand, for the photon we have

$$q^2 = (p_3 - p_4)^2$$

$$= p_3^2 + p_4^2 - 2(\mathbf{p}_3 \cdot \mathbf{p}_4 - E_3 E_4)$$

$$= -2m_e^2 - 2p_3 p_4 \cos\theta + 2E_3 E_4$$

where $p_3 = |\mathbf{p}_3|$, $p_4 = |\mathbf{p}_4|$ and θ is the angle the electron scatters through. If the electron is very relativistic, then $E_3, E_4 \gg m_e$, $E_3 = p_3$, $E_4 = p_4$ and we have

$$q^2 = 2E_3 E_4 (1 - \cos\theta)$$

$$= 4E_3 E_4 \sin^2(\theta/2) > 0 \qquad (6\text{-}8)$$

which is a *positive* quantity. This is to be compared with a real photon which, since it has zero mass, has $q^2 = 0$. The convention is to call states with

ELECTROMAGNETIC STRUCTURE OF NUCLEONS 101

$q^2 < 0$ timelike (massive particles may be real)

$q^2 = 0$ lightlike (massless particles real)

$q^2 > 0$ spacelike (all particles virtual)   (6-9)

The exchanged photon in *e-p* scattering is spacelike, and therefore *must* be virtual, since a real photon is lightlike.

With these kinematic preliminaries settled, let us now turn to the structure of the proton. A simple way to represent its extended structure is to replace the product of wave functions

$$\int \exp(ip_2 . x) \exp(-ip_1 . x) \exp(-iqx) d^4x \qquad (6\text{-}10)$$
    ↑   ↑   ↑
 final*p*   initial*p*   γ

in which the photon is absorbed on a point proton, by, say

$$\iint \exp(ip_2 . x) \exp(-ip_1 . x) F(x' - x) \exp(-iq . x') d^4x d^4x', \qquad (6\text{-}11)$$

in which the photon is absorbed by the virtual meson cloud at x', the meson cloud having been emitted from the proton at x. $F(x - x')$ is a measure of the *charge distribution*. A *point* is represented by the Dirac delta function $F(x - x') = \delta(x - x')$, and in this case (11) reduces again to (10). Equation (11) may be written

$$\int e^{i(p_2 - p_1 - q) . x} F(q^2) d^4x \qquad (6\text{-}12)$$

where

$$F(q^2) = \int e^{-iq . y} F(y) d^4y \qquad (6\text{-}13)$$

is the (4-dimensional) Fourier transform of the charge distribution $F(x - x')$, and is called the *form factor* of the proton. It is to be noted that the integrals in

Figure 6.4 Feynman diagram for $e^- + p \to e^- + p$

(10) to (13) are 4 dimensional, as they should be, to be properly relativistic. A three dimensional charge distribution is ambiguous, since for example a distribution which is spherically symmetric in one frame is not spherically symmetric in another. This means, however, that $F(x - x')$ does not have a simple interpretation as a space charge distribution, except in the nonrelativistic limit where $q_4 = i$(energy transfer) is small (i.e. when there is only a small proton recoil).

It is useful to calculate the form factor for the case of a Yukawa potential, using the nonrelativistic version of (13)

$$F(q^2) = \int e^{-i\mathbf{q}\cdot\mathbf{r}} \rho(\mathbf{r}) d^3 r \qquad (6\text{-}14)$$

where the (spherically symmetric) charge distribution $\rho(\mathbf{r})$ is

$$\rho(r) = -\frac{m^2}{4\pi} \frac{e^{-mr}}{r}$$

where m is the mass of the Yukawa meson. By choosing $\mathbf{q}$ along the z axis we get

$$F(q^2) = \frac{m^2}{4\pi} \iiint e^{-iqr\cos\theta} \frac{e^{-mr}}{r} r^2 dr d\cos\theta d\varphi$$

i.e.

$$F(q^2) = \frac{m^2}{q^2 + m^2} = \frac{1}{1 + \frac{q^2}{m^2}} \qquad (6\text{-}15)$$

EXERCISE   Prove equation (15).

The form factor (15) describes the structure of the proton as a function of q^2. It has a *simple pole* at $q^2 = -m^2$. This is a timelike value, whereas in the scattering experiment q^2 is spacelike, so the pole lies outside the range of q^2 covered in the experiment.

A *point charge* is represented, as mentioned above, by the Dirac delta function $F(\mathbf{x} - \mathbf{x}') = \delta(\mathbf{x} - \mathbf{x}')$. It is easily seen, from (13), that the corresponding form factor is

(Point charge) $F(q^2) = 1;$ \qquad (6\text{-}16)

a constant, as expected. These charge distributions and form factors appear in Table 1.

ELECTROMAGNETIC STRUCTURE OF NUCLEONS

Table 6.1

Charge distributions, form factors and mean square charge radii for various potentials

Charge distribution $\rho(r)$	Form factor $F(q^2)$	Size $\langle r^2 \rangle$
Point $\delta(\mathbf{r} - \mathbf{r}')$	Constant 1	0
Yukawa $\dfrac{m^2}{4\pi}\dfrac{e^{-mr}}{r}$	Simple pole $\dfrac{1}{1 + \dfrac{q^2}{m^2}}$	$\dfrac{6}{m^2}$
Exponential $\dfrac{m^3}{8\pi} e^{-mr}$	Dipole $\left(\dfrac{1}{1 + \dfrac{q^2}{m^2}}\right)^2$	$\dfrac{12}{m^2}$

In the non-relativistic limit, for small q^2, we may define a *size* associated with the charge distribution. This is done as follows. Performing the θ and φ integrations in (14) it is easily seen that

$$F(q^2) = \int \rho(r) \frac{\sin qr}{qr} 4\pi r^2 dr \qquad (6\text{-}17)$$

if $\rho(r)$ is spherically symmetric. For small q^2 we may put

$$\sin qr = qr - \frac{1}{3!}(qr)^3 + \ldots$$

and keep only the first two terms. The total charge Q is given by

$$Q = \int \rho(r) 4\pi r^2 dr.$$

Defining the *mean-square charge radius* $\langle r^2 \rangle$ by

$$\langle r^2 \rangle = \frac{1}{Q} \int \rho(r) 4\pi r^4 dr \qquad (6\text{-}18)$$

we then have

$$F(q^2) \approx Q(1 - \frac{1}{6}\langle r^2 \rangle q^2). \qquad (6\text{-}19)$$

For the Yukawa potential, it follows by expanding (15) in q^2 and comparing with (19) that $\langle r^2 \rangle = 6/m^2$: the heavier the Yukawa meson, the smaller the extent of the charge distribution.

6.3 PROTON AND NEUTRON FORM FACTORS

The pion clouds give structure to both the charge and magnetic moment distributions of the nucleon. These, however, are not simply related, as is clear from (1) to (3). To express this we introduce *separate* form factors for the charge and magnetic moment distributions of both proton and neutron. This involves four functions in all

$G_E^p(q^2)$: electric form factor of p

$G_E^n(q^2)$: electric form factor of n

$G_M^p(q^2)$: magnetic form factor of p

$G_M^n(q^2)$: magnetic form factor of n. (6-20)

The values (3) of the "actual" charge and magnetic moment would be the values "seen" by a photon of very long wavelength, i.e. with $q \to 0$, so on comparing (3) and (20)

$G_E^p(0) = 1, \qquad G_E^n(0) = 0$

$G_M^p(0) = 2.79, \qquad G_M^n(0) = -1.91$ (6-21)

where the units for G_M are nuclear magnetons.

The differential scattering cross section for one photon exchange, as in Figure 4, is given by the *Rosenbluth formula*

$$\frac{d\sigma}{d\Omega} = \left(\frac{d\sigma}{d\Omega}\right)_{\text{point}} \left\{ \frac{(G_E^p)^2 + \dfrac{q^2}{4M^2} \dfrac{(G_M^p)^2}{\mu_N^2}}{1 + \dfrac{q^2}{4M^2}} + \frac{q^2}{2M^2} \frac{(G_M^p)^2}{\mu_N^2} \tan^2(\theta/2) \right\} \quad (6\text{-}22)$$

where M is the proton mass, θ is the electron scattering angle and $(d\sigma/d\Omega)_{\text{point}}$ is the differential cross section for the scattering of point particles, called Mott scattering

ELECTROMAGNETIC STRUCTURE OF NUCLEONS

$$\left(\frac{d\sigma}{d\Omega}\right)_{point} = \left(\frac{d\sigma}{d\Omega}\right)_{Mott} = \frac{a^2}{4E_4^2} \cdot \frac{\cos^2(\theta/2)}{\sin^4(\theta/2)} \cdot \frac{1}{1 + \left(\frac{2E_4}{M}\right)\sin^2(\theta/2)} \quad (6\text{-}23)$$

The second factor in (22) describes how the scattering is affected by the structure of the nucleon. The effect is most noticeable at large q^2: in general terms, if we substitute the form factors $F(q^2)$ from Table 1 for the Gs in (22), then we see that

$$\text{Yukawa:} \quad \frac{d\sigma}{d\Omega} = \left(\frac{d\sigma}{d\Omega}\right)_{point} \cdot \frac{1}{q^4} \text{ as } q^2 \to \infty$$

$$\text{Exponential:} \quad \frac{d\sigma}{d\Omega} = \left(\frac{d\sigma}{d\Omega}\right)_{point} \cdot \frac{1}{q^8} \text{ as } q^2 \to \infty \quad (6\text{-}24)$$

and the cross section in both cases decreases rapidly compared with the scattering from a point particle.

To find the nucleon form factor, fix q^2 and plot

$$\frac{d\sigma}{d\Omega} \left[\frac{a^2}{4E_4^2} \frac{\cos^2(\theta/2)}{\sin^4(\theta/2)} \frac{1}{1 + (2E_4/M)\sin^2(\theta/2)}\right]^{-1}$$

against $\tan^2(\theta/2)$. According to (22) and (23) a straight line should result, whose gradient and intercept give $G_E^p(q^2)$ and $G_M^p(q^2)$. By repeating this procedure at different values of q^2, the form factors may be determined as functions of q^2. The neutron form factors are found from electron-deuteron scattering, subtracting the electron-proton scattering contribution, and making a small correction for the binding.

The behaviour of $G_M^p(q^2)$ and $G_E^p(q^2)$ is shown in Figure 5. $G_{E,M}^p(q^2)$ are $G_M^n(q^2)$ are found to have exactly the same dependence on q^2; this simple rule is known as the *scaling law*

$$G_E^p(q^2) = \frac{G_M^p(q^2)}{2.79} = \frac{G_M^n(q^2)}{-1.91} = G(q^2). \quad (6\text{-}25)$$

In addition

$$G_E^n(q^2) = 0. \quad (6\text{-}26)$$

The single form factor $G(q^2)$ is well described by the *dipole fit*

$$G(q^2) = \left(\frac{1}{1+q^2/0.71}\right)^2 \tag{6-27}$$

where q^2 is measured in $(\text{GeV}/c)^2$. This is seen to correspond, from Table 1, to an exponential, rather than to a Yukawa charge distribution. This is rather surprising, and its significance is not understood. It is quite interesting that in the late 1950s, when experiments were only done at low q^2, it was found possible to fit $G(q^2)$ by a sum of simple poles

$$\text{low } q^2: G(q^2) = \sum_i \frac{g_i}{1 + \frac{q^2}{M_i^2}} \tag{6-28}$$

indicating Yukawa-type mesons. In fact in 1959 this fit was made by Fraser and Fulco, who predicted mesons at the relevant masses $q^2 = -M_i^2$ (*timelike*). These mesons were interpreted as being $\pi\pi$ resonances, and correspond for example to a simple-minded interpretation of the second diagram on the right hand side of Figure 3 – see Figure 6. A π^+ travelling forwards in time is equivalent, according to Feynman, to a π^- travelling backwards in time, so we can draw (b) as equivalent to (a). The two pions in (b) are then taken to be in a resonant state, as in

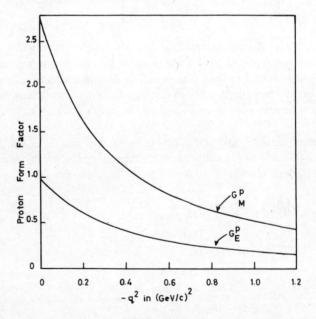

Figure 6.5

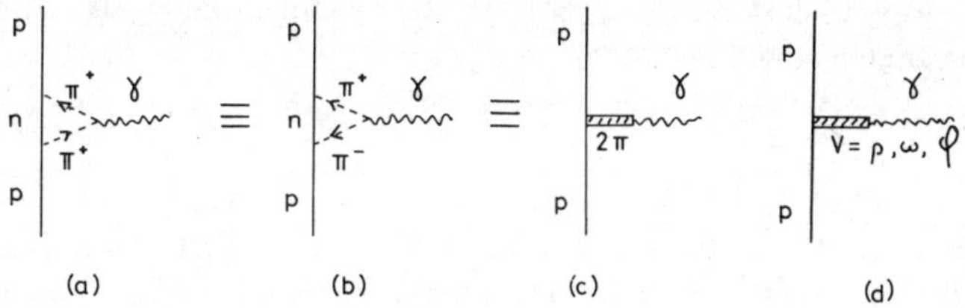

ELECTROMAGNETIC STRUCTURE OF NUCLEONS

[Figure 6.6 diagrams: (a), (b), (c), (d)]

Figure 6.6

(c). This $\pi\pi$ resonance must have $J^P = 1^-$, the same as the photon. Frazer and Fulco, and Nambu, predicted both the ρ and ω mesons this way, though when they were discovered, they were found to have a higher mass than predicted.

Finally, two remarks may be made about the dipole fit (27). Firstly, as seen in (24), the differential cross section decreases very rapidly as q^2 increases. Secondly, the exponential charge distribution which corresponds to a dipole form factor (see Table 1) has the property that

$$\rho(r) = \frac{m^3}{8} e^{-mr}: \quad \rho(r) \to \text{constant as } r \to 0$$

indicating that *there is no hard core in the nucleon*. (A Yukawa charge distribution, in contrast, does not have this property.) Note that this does not imply that there are no hard "seeds" in the proton, only that there is not an accumulation of them at the centre.

The proton, then, is not like a plum, with a stone in the mdidle. The question remains, is it like jelly, or is it like a pomegranate, with seeds in? To answer this we turn to inelastic scattering.

6.4 INELASTIC ELECTRON-PROTON SCATTERING

In the atomic scattering of electrons, besides the elastic process which leaves the atom in its ground state, there are inelastic processes in which the atom is excited to a discrete level, and, beyond that, there is a continuum of inelastic processes ionising the atom. Similarly in electron-nucleus scattering, the nucleus may be left in its ground state (elastic scattering), or, inelastically, be excited to a discrete level, or indeed be disintegrated and eject individual nucleons, analogous to atomic ionisation. In a similar way to atomic and nuclear scattering, besides elastic electron-proton scattering, there is inelastic scattering in which the proton is either excited to a discrete level – a resonance – or beyond that to the continuum where *pionization,* as it is called, sets in.

Let us first look at the kinematics of *e-p* scattering, as represented in Figure 7. The initial and final electron 4-momenta are p and p', the initial proton momentum is P and the final state, whatever it be, has 4-momentum P_f. It is useful to define the variable ν by

$$M\nu = -q.P \tag{6-29}$$

where M is the proton mass. In the frame in which the target proton is at rest, $P = (0, 0, 0, iM)$, so $\nu M = -q_4(iM) = (E - E')M$, where E and E' are the initial and final energies of the electron in this same frame. So

$$\nu = E - E' = \text{energy-loss in target rest-frame.} \tag{6-30}$$

Now $q + P = P_f$ so the final state has mass M_f given by

$$M_f^2 = -P_f^2 = -(q + P)^2$$

$$= -q^2 - P^2 - 2P.q$$

$$= -q^2 + M^2 + 2M\nu,$$

$$M_f^2 = M^2 - q^2 + 2M\nu \tag{6-31}$$

where (29) has been used. In elastic scattering $M_f = M$, so (31) gives

$$\text{Elastic scattering: } q^2 = 2M\nu \tag{6-32}$$

If a resonance is excited then $M_f = M_{res}$ and

$$\text{Resonance excitation: } q^2 = 2M\nu + M^2 - M_{res}^2. \tag{6-33}$$

Finally, when the proton is excited beyond the resonance region into the continuum (so-called *deep inelastic scattering*), we have

$$\text{Deep inelastic scattering: } q^2 = 2M\nu + M^2 - \underset{\underset{\text{continuum}}{\uparrow}}{M_f^2} \tag{6-34}$$

Figure 6.7 Kinematics of the process $e^- + p \to e^- + \text{anything}$

ELECTROMAGNETIC STRUCTURE OF NUCLEONS

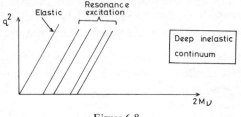

Figure 6.8

These three kinematic regions are shown in Figure 8.

The important point about the inelastic scattering experiments is that once we are in the continuum in Figure 8, the scattering process is effectively

$$e^- + p \to e^- + \text{any hadrons}.$$

Moreover, if in the experiment only the scattered electron is observed, and not the final hadron state, then the measured cross section is the sum of the cross sections for producing many different hadronic final states. Feynman calls such cross sections *inclusive* cross sections. It is then obvious from (34) that since M_f^2 is the continuum and is not measured, it takes on all values and q^2 and ν are *independent variables*. In elastic scattering and resonance excitation, from (32) and (33), this is not true. Hence in elastic scattering there is an extra kinematic degree of freedom. It is this which enables more information to be extracted from inelastic than from elastic scattering experiments. In our nonrelativistic picture, elastic scattering gives the charge distribution in space (**r** = Fourier transform of **q**), but inelastic scattering gives the *instantaneous* charge distribution, that is, the charge distribution in space and time (**r**, t = Fourier transforms of **q**, E).

Equation (22) for elastic scattering may be written

$$\left(\frac{d\sigma}{dq^2}\right)_{\text{elastic}} = \left(\frac{d\sigma}{dq^2}\right)_{\text{point}} [A(q^2) + B(q^2)\tan^2(\theta)]$$

where A and B are combinations of the form factors G_E and G_M. In inelastic scattering this is generalised simply to take account of the fact that there are now two variables q^2 and ν, instead of just q^2, so we put

$$\frac{d^2\sigma}{dq^2 d\nu} = \frac{4\pi a^2}{q^4} \cdot \frac{E'}{E} [2W_1(q^2, \nu)\sin^2(\theta/2) + W_2(q^2, \nu)\cos^2(\theta/2)] \quad (6\text{-}35)$$

where W_1 and W_2 are called *structure factors*. They are the inelastic analogue of form factors. By performing the experiment at different angles W_1 and W_2 can be separated. At small angles, however, W_2 dominates, and the results presented in Figure 9 for an experiment performed at 6°, show the variation of W_2 with ν, for different values of q^2. The following features may be noted:

a) At small ν the bumps in W_2 correspond to the elastic form factor, and to excitation of discrete resonances. When $\nu \gtrsim 3$ GeV, the curve becomes smooth, corresponding to "pionisation", i.e. the final state is in a continuum.

b) At fixed q^2, if $d^2\sigma/dq^2 d\nu$ is integrated over ν, the long tail on the curve results in a *large* value for $(d^2\sigma/dq^2)_{\text{inelastic}}$, comparable with the Mott cross section from a point proton. This is in sharp contrast with the elastic scattering cross section (24), which decreases rapidly with increasing q^2.

c) Experimental points on the graph of W_2 against ν fall on the *same curve* for all values of q^2, if $\nu \gtrsim 4$ GeV. In other words, W_2 becomes a function of the ratio ν/q^2 only, rather than of ν and q^2 separately, as would be expected *a priori*. Defining $\omega = M\nu/q^2$, ω is dimensionless, and therefore *scale invariant*, so we see that W_2 becomes scale invariant at high ν.

Feature (b) strongly suggests that the scattering is from *pointlike constituents* of the nucleon. The situation may be compared with Rutherford's discovery that a-particles scattering from atoms only interact with the point (at least almost point) nucleus. Feynman has constructed a model in which the nucleon is "made out of" a loosely bound collection of pointlike constituents, which he calls *partons*.[†] In deep inelastic scattering, the photon interacts with only *one* of these partons. By contrast, elastic scattering is coherent scattering from the cloud of partons, whose spatial distribution gives (by Fourier transform) the strong q^2 dependence of the form factors.

It can be shown that the parton model explains the observed scaling in (c). Actually, the scaling property had been predicted by Bjorken on other theoretical grounds. The experiments, nevertheless, did come as a surprise, and have given birth to much theoretical activity which is still alive. Feynman's theory of partons is only one theory among many − though it is the most picturesque.

It is tempting to sepculate that partons may be quarks, the possible constituent of hadrons suggested by unitary symmetry (see Chapter 11). Attempts to explain the data by assuming that partons are quarks have met with some degree of success. But in any case it appears that the deep inelastic scattering experiments may have revealed the existence of "seeds" in the proton − that it is like a pomegranate, not like jelly.

[†] For an extensive discussion of the parton model, see R. P. Feynman, *Photon-Hadron Interactions*, Benjamin, 1972

ELECTROMAGNETIC STRUCTURE OF NUCLEONS 111

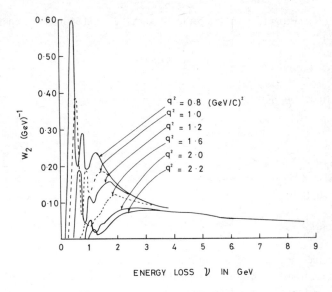

Figure 6.9 A plot of $W_2(q^2, \nu)$ against $\nu = E - E'$ for various values of q^2. (After Panofsky, 14th International Conference on High Energy Physics, Vienna, 1968. CERN, Geneva, 1968)

6.5 EXPERIMENTS WITH TIMELIKE PHOTONS

In both processes discussed so far, the photons were spacelike, $q^2 > 0$. This, as we saw in (8), was a consequence of the fact that $q = p_3 - p_4$, the *difference* between two timelike momenta. One of the tenets of particle physics, however, is that the dynamics of a particular process does not depend in an essential way on the particular values of energy and momentum. Thus, for example, the dynamical features of elastic *e-p* scattering, where $q^2 > 0$, should also be present in the process where $q^2 < 0$, where the photon is timelike. This is the case in the reaction

$$e^- + e^+ \to \bar{p} + p \tag{6-36}$$

whose Feynman diagram is shown in Figure 10. This reaction is electron-positron annihilation. It obviously requires an extremely high energy to produce p and $\bar{p}$.

In the last few years experiments of this type have become possible using *colliding beams* of e^- and e^+ which circulate in storage rings in opposite directions at high energy, colliding generally once or twice every revolution. The first experiments on colliding beams were performed in 1967 and 1968 at Novosibirsk (Siberia) and Orsay (France). The great kinematic advantage of colliding beams is that the centre of mass energy is the laboratory energy (in conventional accelerators with a sitting target it is a lot less), so that e^- and e^+ have the energy-momenta

$$p_3 = (0, 0, p, iE)$$

$$p_4 = (0, 0, -p, iE)$$

and

$$q = p_3 + p_4$$

$$= (0, 0, 0, 2iE)$$

$$q^2 = -4E^2. \tag{6-37}$$

q^2 is timelike, as expected, and in this process q is the *energy* rather than the momentum transfer. At low energies there is not enough energy to produce $p\bar{p}$, but there is enough to produce pions

$$e^+ + e^- \to 2\pi, 3\pi. \tag{6-38}$$

Returning now to the process of elastic electron-nucleon scattering, if there is any truth in the idea that photons behave like the vector mesons ρ, ω and φ when interacting with hadrons, as suggested by the prediction of ρ and ω from the nucleon form factors, according to the Feynman diagram of Figure 6, then we should expect that the timelike photon in the process (38) should also "turn into" a vector meson. This is illustrated in Figure 11. If this diagram represents what happens in the process (38), then we should expect to see the 2π, for example, *peaked* at the ρ mass of 765 MeV, corresponding to

$$e^+ + e^- \to \rho \to \pi^+ + \pi^-.$$

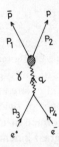

Figure 6.10 Feynman diagram for $e^+ + e^- \to p + \bar{p}$

ELECTROMAGNETIC STRUCTURE OF NUCLEONS

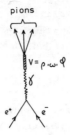

Figure 6.11 Feynman diagram for $e^+ + e^- \to n\pi$, mediated by vector mesons ρ, ω and φ

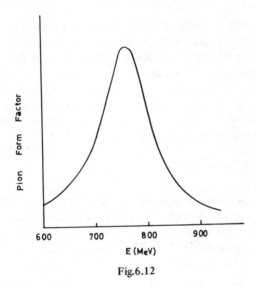

Fig.6.12

This has indeed been seen at Orsay and Novosibirsk. The Orsay result is shown in Figure 12. The reactions

$$e^+ + e^- \to \omega \to \pi^+ + \pi^- + \pi^0$$

$$e^+ + e^- \to K + \bar{K}$$

have also been observed in other colliding beam experiments.

These experiments provided confirmation of the ideas developed from nucleon form factors. The similarity of the processes, in fact, goes even further. Just as the sum of simple poles (28), from which the vector mesons were predicted, only holds at low q^2, so also recent colliding beam experiments at Frascati (Italy) indicate that at high q^2 the production of pions is greater than the diagram of Figure 11 would imply. In both cases the vector mesons are relevant, but do not provide a complete description of the way in which photons interact with hadrons.

At higher energies the colliding rings of e^- and e^+ are able to produce much more than simply pions. Of particular interest is the cross section for

$$e^+ + e^- \to \text{any hadrons} \tag{6-39}$$

summed over all hadron channels. This process has the Feynman graph of Figure 13, and is simply the timelike-γ version of

$$e^- + p \to e^- + \text{any hadrons} \tag{6-40}$$

where γ is spacelike, according to the Feynman diagram of Figure 7. Just as the experiments on (40) indicated the possible presence of pointlike hadronic constituents, so it may be expected that the colliding beam experiments (39) should furnish extra evidence on this question. If partons exist, then the energy (q^2) dependence of (39) should be $\sigma \propto 1/E^2$ rather than a more rapid fall-off characteristic of q^2-dependent elastic form factors. In fact, experiments at Frascati reported by Drell (see Further Reading), in the energy interval $1.4 \text{ GeV} \leqslant 2E \leqslant 2.4 \text{ GeV}$, give a total cross section (averaged over energy)

$$\sigma(e\bar{e} \to \text{hadrons}) \approx 30 \pm 10 \text{ nb} \tag{6-41}$$

whose energy dependence is "not incompatible" with a $1/E^2$ fall-off. (This dependence of σ on E follows easily from the scale invariance discussed above. If σ only depends on the kinematics, then since it is a (length)^2, and length = $c\hbar$ (energy)$^{-1}$, we must have $\sigma \propto E^{-2}$. This is merely dimensional analysis with free use of c and $\hbar$.) Moreover, the cross section for $e\bar{e}$ annihilation to pointlike muon pairs is

$$\sigma(e\bar{e} \to \mu\bar{\mu}) \approx 21 nb/E_{\text{GeV}}^2,$$

which is comparable with (41).

Figure 6.13

FURTHER READING

Drell, S. D., "Electromagnetic Interactions with emphasis on Colliding Rings, Partons and the Light Cone", *Proc. Amsterdam Conf. on Elementary Particles 1971*, (A. G. Tenner and M. J. G. Veltman, eds.) North-Holland, 1972.

Griffy, T. A. and L. I. Schiff, "Electromagnetic Form Factors" in *High Energy Physics*, (E. H. S. Burhop, ed.) vol I, Academic Press, 1967.

Jackiw, R., "Introducing Scale Symmetry", *Physics Today*, January 1972, p.23.

Kendall, H. W. and W. K. Panofsky, "The Structure of the Proton and the Neutron", *Scientific American*, **224**, 60, (1971)

Wilson, R., "Form Factors of Elementary Particles", *Physics Today*, January 1969, p.47.

CHAPTER 7

Resonances

AS BIGGER accelerators enable higher energies to be reached, more and more resonances are discovered. A resonance is a hadron state with well-defined spin, parity, isospin and strangeness. The dynamics of the strong interactions by which these resonances are produced is one of the major unsolved problems in the physics of elementary particles. It is not one which is considered in this book,[†] though in the last few years there have been some very interesting developments connected with theory of Regge poles. This chapter presents, far more mundanely, a simple phenomenological survey of a few resonances, which are taken to be fairly typical. Figures 2 and 4 show the baryon and meson resonances up to about 1900 MeV. The chapter ends with a few remarks about whether resonances are as elementary as the "stable" hadrons.

7.1 AN EXAMPLE: Δ (1236)

This state, sometimes denoted N^*_{33} (33 stands for spin 3/2 and isospin 3/2), was the first resonance to be discovered. It was found in 1952 by Fermi at Chicago as a large peak in the pion nucleon scattering cross section, at a pion energy of about 195 MeV. No-one then realised what a deluge of resonances was in store.

The mass is deduced directly from the fact that the resonance peak occurs at a pion laboratory kinetic energy T_π of 195 MeV; this is the first peak in Figure 1. It corresponds to

$$\pi + p \to \Delta \to \pi + p. \tag{7-1}$$

Since the rest-mass m_π = 140 MeV/c², the total pion energy is 195 + 140 = 335 MeV, and its momentum p_π in the laboratory frame is

$$p_\pi = \sqrt{E_\pi^2 - m_\pi^2} = \sqrt{335^2 - 140^2} = 304 \text{ MeV/c}.$$

---

† For an introduction to strong interaction dynamics, see the books by R. Omnès and D. H. Perkins in 'Further Reading' at the end of this chapter.

RESONANCES

In this same lab frame the proton is at rest, so defining the z axis to be the direction along which the pion is moving, the four-momenta are, in MeV

$P_\pi = (0, 0, 304, 335i)$

$P_p = (0, 0, 0, 938i)$

giving a total energy-momentum at resonance of

$P_\Delta = P_\pi + P_p = (0, 0, 304, 1273i),$

corresponding to

$M_\Delta = \sqrt{1273^2 - 304^2} = 1236 \text{ MeV}/c^2.$

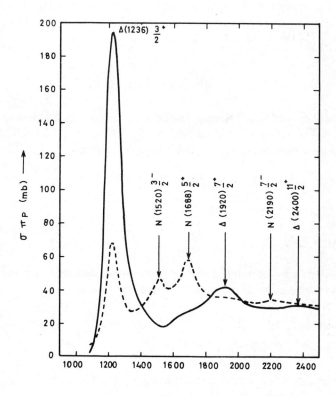

Figure 7.1 Total cross sections for $\pi^+ p$ (solid line) and $\pi^- p$ (dashed line) scattering

The isospin must be $\frac{1}{2}$ or $\frac{3}{2}$, since I is conserved in the production reaction. From the fact that the resonance peak occurs in the $\pi^+ p$ cross section (where $I_3 = \frac{3}{2}$) as well as in the $\pi^- p$ cross section (where $I_3 = -\frac{1}{2}$) it is clear that $I = \frac{3}{2}$, and the Gell-Mann–Nishijima relation gives the charge states as

$$\Delta: \Delta^{++}, \Delta^+, \Delta^0, \Delta^-. \quad (I = \tfrac{3}{2})$$

Finally, let us consider the spin and parity. The easiest way to proceed is to assume that Δ has spin $\frac{3}{2}$, which it has. It decays strongly into N ($s = \frac{1}{2}$) and π ($s = 0$) with orbital angular momentum $l = 1$ or 2, by the vector addition rule. The parity of Δ is

$$\epsilon(\Delta) = \epsilon(N)\epsilon(\pi)(-1)^l = (-1)^{l+1} \tag{7-2}$$

and depends on l. Denote the spin of Δ by j, and assume arbitrarily that $j_z = \frac{1}{2}$. We must now express the state (j, j_z) in terms of the states (l, m) and (s, s_z) where l is the relative angular momentum of the pion and nucleon and s is the proton spin. Since $(j, j_z) = (\frac{3}{2}, \frac{1}{2})$ and $s = \frac{1}{2}$, then
(a) if $l = 1$,

$$\psi_j(\tfrac{3}{2}, \tfrac{1}{2}) = c_1 \psi_l(1, 1)\psi_s(\tfrac{1}{2}, -\tfrac{1}{2}) + c_2 \psi_l(1, 0)\psi_s(\tfrac{1}{2}, \tfrac{1}{2})$$

where c_1 and c_2 are Clebsch-Gordon coefficients. By making the substitutions (orbital $l = 1$) $\rightarrow$ (pion $I = 1$) and (nucleon $s = \frac{1}{2}$) $\rightarrow$ (nucleon $I = \frac{1}{2}$), this equation has its isospin analogue in equation (3A-20), so we may put

$$\psi_j(\tfrac{3}{2}, \tfrac{1}{2}) = \sqrt{\tfrac{1}{3}}\psi_l(1, 1)\psi_s(\tfrac{1}{2}, -\tfrac{1}{2}) + \sqrt{\tfrac{2}{3}}\psi_l(1, 0)\psi_s(\tfrac{1}{2}, \tfrac{1}{2}).$$

$\psi_l(l, m)$, the orbital wave function, is

$$\psi_l(l, m) = Y^l_m(\theta, \varphi),$$

the associated Legendre polynomial. $\psi_j(\frac{3}{2}, \frac{1}{2})$ is now expressed as a function of θ and φ, and the intensity distribution of the pions is

$$I(\theta) = \psi_j^*(\tfrac{3}{2}, \tfrac{1}{2})\psi_j(\tfrac{3}{2}, \tfrac{1}{2})$$
$$= \tfrac{1}{3}|Y^1_1|^2 + \tfrac{2}{3}|Y^1_0|^2.$$

The interference (cross) term vanishes because Y^1_1 and Y^1_0, as well as $\psi_s(\frac{1}{2}, -\frac{1}{2})$ and $\psi_s(\frac{1}{2}, \frac{1}{2})$, are orthogonal. Substituting

$$Y^1_1 = -\sqrt{\tfrac{3}{8\pi}}\sin\theta\, e^{i\varphi}, \quad Y^1_0 = \sqrt{\tfrac{3}{4\pi}}\cos\theta$$

gives

$$I(\theta) \propto \sin^2\theta + 4\cos^2\theta = 1 + 3\cos^2\theta. \tag{7-3}$$

(b) *if l = 2*,

$$\psi_j(3/2, 1/2) = \sqrt{3/5}\, Y_1^2 \psi_s(1/2, -1/2) - \sqrt{2/5}\, Y_0^2 \psi_s(1/2, 1/2).$$

The Clebsch–Gordon coefficients $\sqrt{3/5}$ and $-\sqrt{2/5}$ are obtained from the "2 × ½" table in Appendix 3A. Since

$$Y_1^2 = -\sqrt{15/8\pi}\,\sin\theta\cos\theta\, e^{i\varphi}, \quad Y_0^2 = \sqrt{5/4\pi}\,(3/2 \cos^2\theta - 1/2),$$

we have for the angular distribution of pions

$$I(\theta) = 5/4\pi\,[3/5 \cdot 3/2 \sin^2\theta\cos^2\theta + 2/5 \cdot 1/4 (9\cos^4\theta - 6\cos^2\theta + 1)] \propto 1 + 3\cos^2\theta. \tag{7-4}$$

Comparison of (3) and (4) shows that *the angular distribution depends only on j,* not on *l*.

EXERCISE   Prove that if a pion-nucleon resonance has $j = 1/2$, then for both $l = 0$ and $l = 1$, the angular distribution is isotropic.

The point of the calculation above is to show that the spin *j* of a pion-nucleon resonance is obtained from the angular distribution of the scattered pions, but this does *not* determine *l*, and hence from (2) does not give the parity of the resonance. *l* is found either from a *phase shift analysis* or by measuring the polarisation of the outgoing nucleon.[†] Both methods yield, in the case of Δ, $l = 1$, so the resonance has even parity: $J^P = 3/2^+$.

7.2   Σ (1385)

This was the first strange resonance to be found, and is often denoted $Y_1^*(1385)$. It was found in the reaction

$$K^- + p \to \Lambda + \pi^+ + \pi^-$$

by observing a peak in the pion energy spectrum. A peak corresponds to the fact that the final consists not of three independent particles but of two, so that the reaction is

---

[†] See, for example, G. Källén, *Elementary Particle Physics*, Addison-Wesley, Chapter 4. This is an excellent account of pion-nucleon scattering, at a slightly more advanced level than in this chapter. Källén uses Dirac's bra-ket notation.

$$K^- + p \to \Sigma(1385) + \pi$$
$$\hookrightarrow \Lambda + \pi \tag{7-5}$$

The mass of $\Sigma(1385)$ can be calculated from the peak pion energy as follows. In the centre of mass system

$$W = E_1 + E_2 + E_3$$

$$0 = \mathbf{p}_1 + \mathbf{p}_2 + \mathbf{p}_3$$

where $(E_1\, p_1)$ etc. are the energies and momenta of the three final state particles. The invariant (mass)^2 of the pair (12) — in our case $(\Lambda\pi)$ — is

$$M_{res}^2 = (E_1 + E_2)^2 - (\mathbf{p}_1 + \mathbf{p}_2)^2$$

$$= (W - E_3)^2 - \mathbf{p}_3^2$$

$$= W^2 + m_3^2 - 2WE_3$$

and is proportional to the energy of the third particle. From the peak pion energy E_3, M_{res} may then be determined.

Processes of the type (5) are called *formation* experiments; the resonance is produced alongside another particle. In contrast, processes of the type (1) are called *production* experiments, where the resonance is produced alone. The distinction is an important one for experimentalists, since the kinematics in formation processes is slightly more involved.

To return to $\Sigma(1385)$, its isospin is clearly $I = 1$, as seen from its decay. Its spin and parity are measured from the angular distribution of the decay products and the polarisation of Λ, and are $J^P = \tfrac{3}{2}^+$; the same as $\Delta(1236)$.

7.3 $\Lambda(1405)$ and $\Lambda(1520)$

These have both been seen as peaks in the kaon spectrum in

$$\pi^- + p \to \Sigma^+ + K^0 + \pi^-$$

indicating that the reaction is, for example

$$\pi^- + p \to \Lambda(1405) + K^0$$
$$\hookrightarrow \Sigma^+ + \pi^-$$

Both $\Lambda(1405)$ and $\Lambda(1520)$ have isospin zero, and $J^P = \frac{1}{2}^-(1405)$, $J^P = \frac{3}{2}^-(1520)$.

7.4 $\Xi(1530)$

The best established baryon resonance with $S = -2$ is $\Xi(1530)$, discovered in 1962 in the reactions

$$K^- + p \to \begin{cases} K^+ + \Xi^- + \pi^0 \\ K^0 + \Xi^- + \pi^+ \\ K^0 + \Xi^0 + \pi^0 \end{cases} \tag{7-6}$$

corresponding to

$$K + p \to \Xi(1530) + K$$
$$\downarrow \Xi + \pi.$$

The assignment $I = \frac{1}{2}$ is deduced from the relative rates of the first two processes in (6).

7.5 $\Omega(1672)$

This particle was predicted by unitary symmetry in 1962, and has $I = 0, S = -3$. It was first found in 1964 as a bubble chamber event, reconstructed to be

$$K^- + p \to \Omega^- + K^+ + K^0$$
$$\downarrow \Xi^0 + \pi^-$$
$$\downarrow \Lambda + \pi^0$$
$$\downarrow \gamma + \gamma$$
$$\downarrow e^+ + e^-$$
$$\downarrow e^+ + e^-$$
$$\downarrow \pi^- + p$$

Note that the first decay, $\Omega \to \Xi + \pi$ violates S and I, and is therefore weak. Ω is stable against strong interaction decay, so is not, strictly speaking, a resonance. It is discussed further in Chapter 10.

7.6 ρ MESON

The first meson resonance to be discovered was the ρ meson in 1961. A peak was found in the spectrum of pion pairs in the reactions

$$\pi^- + p \begin{cases} \to \pi^+ + \pi^- + n \\ \to \pi^- + \pi^0 + p. \end{cases}$$

The mass is 765 MeV. The same peak is found in the reaction

$$\pi^+ + p \to \pi^+ + \pi^0 + p,$$

but not in $\pi^+ p \to \pi^+ \pi^+ n$, indicating that $I = 1$. Two pions with odd isospin must also have odd relative angular momentum, so $J = 1, 3, \ldots$. The angular distribution of the pions is of the form

$$I(\theta) = A + B\cos\theta + C\cos^2\theta,$$

characteristic of an $l = 1$ state, therefore indicating that $J = 1$. The parity is clearly $(-1)^2(-1)^J = -1$, so $J^P = 1^-$. Mesons with this spin and parity (the same as the photon) are called vector mesons.

It will be recalled from the discussion of form factors in the last chapter, that the existence of the ρ meson was originally predicted from the nucleon form factors. Its discovery, however, was in a pure strong interaction process.

7.7 ω MESON

Another important vector meson is ω which was also discovered in 1961 as a peak in the $(\pi^+\pi^-\pi^0)$ spectrum in the annihilation reaction

$$p + \bar{p} \to \pi^+ + \pi^+ + \pi^0 + \pi^- + \pi^-.$$

No peak is found in the spectra of $(\pi^+\pi^+\pi^-)$, $(\pi^+\pi^+\pi^-)$, $(\pi^+\pi^+\pi^0)$ or $(\pi^-\pi^-\pi^0)$, so ω has $I = 0$.

The best way to construct the $I = 0$ state of 3π is to consider 3π $(I = 0)$ as a combination of π $(I = 1)$ and 2π $(I = 1)$, whose latter wave function can be constructed from the 1×1 table in Appendix 3A:

$$\Psi(3\pi; 0, 0) = \frac{1}{\sqrt{3}} [\varphi_{\pi^+}\Phi(2\pi; 1, -1) + \varphi_{\pi^-}\Phi(2\pi; 1, 1) - \varphi_{\pi^0}\Phi(2\pi; 1, 0)]$$

$$= \frac{1}{\sqrt{6}}(\varphi_{\pi+}\;\varphi_{\pi 0}\varphi_{\pi--} - \varphi_{\pi+}\;\varphi_{\pi-}\varphi_{\pi 0} + \varphi_{\pi-}\varphi_{\pi+}\varphi_{\pi 0} - \varphi_{\pi-}\varphi_{\pi 0}\varphi_{\pi+} +$$

$$+ \varphi_{\pi 0}\varphi_{\pi-}\varphi_{\pi+} - \varphi_{\pi 0}\varphi_{\pi+}\varphi_{\pi-}).$$

This isospin wave function is antisymmetric under exchange of any two pions. Since pions are bosons, this means their *configuration space wave function must also be antisymmetric*. In other words, the decay amplitude $M(p_i, E_i)$, which is a function of the energies and momenta of the pions (labelled by i), must be antisymmetric in p_i and E_i. Since M describes the transition between ω and 3π parity conservation gives

$$\varphi_\omega^* M(\mathbf{p}_i, E_i)\Psi_{3\pi} = (-1)^3 \epsilon(\omega)\varphi_\omega^* M(-\mathbf{p}_i, E_i)\Psi_{3\pi} \tag{7-7}$$

where $\epsilon(\omega)$ is the parity of ω.

We are now in a position to discuss the possible spin-parity assignments for ω. First, 0^+ is impossible by conservation of angular momentum and parity; for if $J(\omega) = 0$, then $L = 0$ for the 3π state whose parity is therefore $(-1)^3(-1)^L = -1$. This is opposite to the positive parity assumed for ω.

Next consider $J^P = 0^-$. In this case, from (7) M is a scalar, and antisymmetric, function of $\mathbf{p}_i$ and E_i. This is achieved by

$$J^P = 0^-: \quad M \propto (E_1 - E_2)(E_2 - E_3)(E_3 - E_1). \tag{7-8}$$

If ω is a 1^+ (pseudovector) meson, M must, from (7), be a vector constructed from E_i and $\mathbf{p}_i$ (and as before, it must be antisymmetric in the suffixes). The simplest such vector is

$$J^P = 1^+: \quad M \propto E_1(\mathbf{p}_2 - \mathbf{p}_3) + E_2(\mathbf{p}_3 - \mathbf{p}_1) + E_3(\mathbf{p}_1 - \mathbf{p}_2). \tag{7-9}$$

On the other hand, if ω is a 1^- (vector) meson, M is pseudovector and we have

$$J^P = 1^-: \quad M \propto \mathbf{p}_1 \times \mathbf{p}_2 + \mathbf{p}_2 \times \mathbf{p}_3 + \mathbf{p}_3 \times \mathbf{p}_1 \propto 3\mathbf{p}_1 \times \mathbf{p}_2 \tag{7-10}$$

since $\mathbf{p}_1 + \mathbf{p}_2 + \mathbf{p}_3 = 0$ in the rest frame of ω. The problem is now to decide which (if any) of the expressions (8), (9) and (10) holds in the actual decay. The relevant analysis was worked out by Dalitz in the decay $K \to 3\pi$ — the important similarity is that the final state involves three particles. The transition rate is as usual given by

$$W = \frac{2\pi}{\hbar}|M|^2 \rho_f,$$

where M is the transition amplitude and ρ_f the density of final states or phase space factor. What Dalitz showed was that *for a three-body final state*, ρ_f depends on the energies of the particles in such a way that[†]

$$\frac{\partial^2 W}{\partial E_1 \partial E_2} = |M|^2.$$

In other words, if each decay event is recorded by a point on a graph of E_1 against E_2, *the density of points is a direct measure of* $|M|^2$. Graphs of this type are called *Dalitz plots*.

It is more usual to plot the points on a symmetric plot with axes E_1, E_2 and E_3 at $120°$. Such a plot for $\omega \to 3\pi$ decay is shown in Figure 3; T_0, T_+ and T_- are the kinetic energies of π^0, π^+ and π^-. Now if M is given by (8), it vanishes along the lines $E_1 = E_2$, $E_2 = E_3$, $E_3 = E_1$, i.e. along the *symmetry axes* of the Dalitz plot, and in this case there would be a depletion of points along these axes. This is not observed, as can be seen from Figure 3. $J^P = 0^-$ is therefore ruled out. Expression (9), in turn, leads to a depletion of points at the centre of the Dalitz plot, where $\mathbf{p}_1 = \mathbf{p}_2 = \mathbf{p}_3$. This is also not observed, ruling out $J^P = 1^+$. Finally, (10) leads to a depletion of points when $\mathbf{p}_1 \parallel \mathbf{p}_2$. This is an extreme kine-

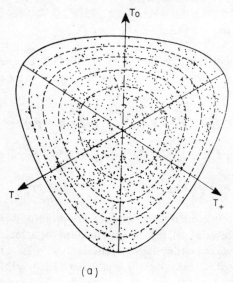

(a)

Figure 7.3 Dalitz plot for $\omega \to 3\pi$ decay. (From C. Alff *et al.*, *Physical Review Letters* 9, 325 (1962))

[†] See, for example, G. Källén, op.cit., pp.196–197.

matic condition and corresponds to the *boundary* of the Dalitz plot. It can be seen in Figure 3 that there is actually a scarcity of points around the boundary. From (10) we conclude that ω has $J^P = 1^-$; it is a vector meson. (Actually, 2^+ gives the same distribution of points as 1^-, but can be excluded by another argument, due to Dalitz.)

7.8 φ MESON

In 1963 a study of the invariant mass of K-$\bar{K}$ in the reactions

$$K^- + p \to \begin{cases} \Lambda + K^+ + K^- \\ \Lambda + K^0 + \bar{K^0} \end{cases}$$

showed a narrow peak at 1020 MeV. This is the φ meson. It has $I = 0$, since the charged K-$\bar{K}$ counterparts in

$$K^- + p \to \Sigma^+ + K^- + K^0$$

do not show a peak. φ decays exclusively into $K\bar{K}$

$$\varphi \to K^+ + K^-$$

and the angular distribution indicates that φ has spin 1. Conservation of parity then gives $P(\varphi) = (-1)^2(-1)^1 = -1$. So φ is a vector (1^-) particle with $I = 0$, $S = 0$; in fact, it has the same quantum numbers as ω. It differs from ω in its decay to $K\bar{K}$, as against ω's decay to 3π.

7.9 η MESON

This was first discovered in 1961 as a peak in the (3π) system in the reaction

$$\pi^+ + d \to p + p + \pi^+ + \pi^- + \pi^0.$$

Its mass is 549 MeV. In this experiment its decay is clearly into 3π

$$\pi^+ + d \to p + p + \eta$$
$$\hookrightarrow \pi^+ + \pi^-\,\pi^0$$

No charged η has ever been observed, so $I = 0$. The most obvious way to find its

126   ELEMENTARY PARTICLES AND SYMMETRIES

spin and parity is to analyze the Dalitz plot for $\eta \to 3\pi$, exactly as was done for $\omega \to 3\pi$. It will be recalled from that analysis that 0^+ is forbidden, and that 0^-, 1^+ and 1^- all give characteristic regions of the Dalitz plot where there should be few events; in all cases, the points on the Dalitz plot give a *non-uniform* distribution. The dalitz plot for $\eta \to 3\pi$, however, gives a *uniform* distribution. What has gone wrong?

The key to this problem is to observe that the main decay modes of η are

$\eta \to 2\gamma$ 39%

$\eta \to 3\pi^0$ or $2\pi^0\gamma$ 31%

$\eta \to \pi^+\pi^-\pi^0$ 25%

$\eta \to \pi^+\pi^-\gamma$ 5%.

$\eta \to 2\gamma$ is an *electromagnetic* decay, so the other modes, being comparable in magnitude, are also electromagnetic. They therefore violate isospin, and the argument followed for ω does not apply. The fact that the Dalitz plot is uniform means that the decay amplitude is constant, so the spin of η is 0. Its parity is therefore negative, i.e. $J^P = 0^-$. It is another pseudoscalar meson, like π and K. It is of interest that Gell-Mann predicted an $I = Y = 0$ pseudoscalar meson in 1961, to fit into a unitary spin supermultiplet with π and K.

7.10 K^* MESON

This appears as a peak in the $K\pi$ mass spectrum in reactions like

$$K^- + p \to \begin{cases} p + \overline{K^0} + \pi^- \\ p + K^- + \pi^0 \end{cases}$$

$$K^+ + p \to \begin{cases} p + K^0 + \pi^+ \\ p + K^+ + \pi^0 \end{cases}.$$

Its mass is 892 MeV. Its isospin (which must be $\tfrac{3}{2}$ or $\tfrac{1}{2}$) is easily determined from a measurement of the width ratio

$$R = \frac{\Gamma(K^{*-} \to \overline{K^0} + \pi^-)}{\Gamma(K^{*-} \to K^- + \pi^0)}.$$

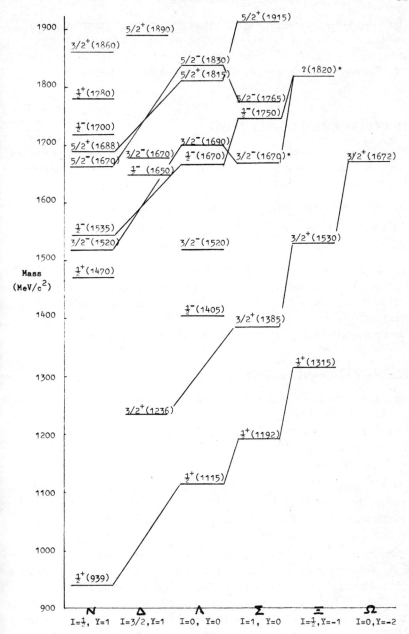

Figure 7.2 Baryon states up to about 1900 MeV. The groupings indicated correspond to SU_3 octets and decuplets (see Chapter 11). *The situation concerning $\Sigma(1670)$ and $\Xi(1820)$ is confused. (Information taken from "Review of Particle Properties", *Reviews of Modern Physics*, 45, no.2, pt. II (supplement), 1973)

As far as isospin is concerned, the πK system is the same as the πN system, so by consulting equations (3A-25) and (3A-26) we see that if K^* has $I = \frac{3}{2}$, then $R = \frac{1}{2}$, whereas if $I = \frac{1}{2}$, then $R = 2$. Experimentally, $R = 1.4 \pm 0.4$, so we conclude that $I = \frac{1}{2}$. The angular distribution of π and K implies that K^* has spin 1, so it is a 1^- meson

7.11 STATES WHICH DO NOT EXIST

There are so many particle states which exist that it becomes equally interesting to enquire if there are any states which "could" exist, but don't. There are states in this category, as may be seen from Figures 2 and 4. In particular

a) Apart from the deuteron d, there are no states with $B = 2$.

b) There are no baryons with $I = \frac{3}{2}$ and $S \neq 0$, e.g. KN resonances.

c) There are no meson states with $I = 2$, e.g. $\pi^+\pi^+$ resonances.

In fact we can make the general observation that as we go to higher mass, we encounter states with higher spin, but not with higher isospin or strangeness. States such as (a), (b) and (c) above are known as *exotic* states, and it is at present not fully understood why they do not exist. The quark model, however, to be discussed in Chapter 11, throws some light on this situation.

7.12 ARE RESONANCES COMPOSITE?

At first sight it is natural to regard resonances as merely composite states of more stable particles, and therefore as less elementary. In the last few years, however, this viewpoint has somewhat lost favour, and resonances are regarded by many physicists as being just as "elementary" as other particles. Several points may be made to support this view:

1) If the mass of Σ^0 were only 6% higher, it would become unstable with respect to the strong decay $\Sigma^0 \rightarrow \Lambda + \pi^0$. Technically speaking, Σ^0 would then be a resonance, but it would be unreasonable in that event to regard it as being intrinsically less elementary than Λ.

2) Many resonances have more than one decay mode. For example, $\Lambda(1520) \rightarrow N\bar{K}$, $\Sigma\pi$ and $\Lambda\pi\pi$; $\varphi \rightarrow K^+K^-$, $K_L K_S$ and 3π. Of which decay products would the state then be a composite? Also, many *virtual* transitions exist, e.g. $\Sigma \rightarrow \Lambda + \pi$, which do not result in decay, because that would be energetically forbidden. An appropriate model for this is to represent resonances as *cavity resona tors*, which at certain frequencies can communicate with certain channels (decay modes). At the wrong frequency, the channel becomes closed; and there are channels below threshold. This model of a resonance is certainly more sophisticated than that of a straightforward composite state. For further details, the reader is referred to

RESONANCES 129

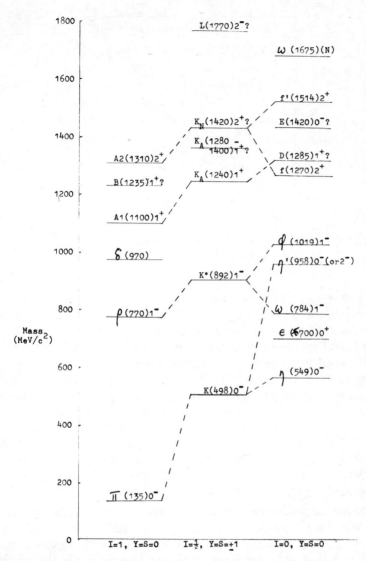

Figure 7.4   Meson states up to about 1800 MeV. The groupings indicated correspond to SU_3 octets and singlets (nonets); see Chapter 11. "N" means parity $P = (-1)^J$ – "normal". "A" means $P = (-1)^{J+1}$ – "abnormal". ? means J^P uncertain. (Information taken from "Review of Particle Properties", *Reviews of Modern Physics*, **45**, no.2, pt II (supplement), 1973)

the article by Chew, Gell-Mann and Rosenfeld (see "Further Reading").

3) If resonances and stable particles are plotted on a graph of J(spin) against (mass)^2, states of the same B, I and S fall on *straight lines*; an example is shown in

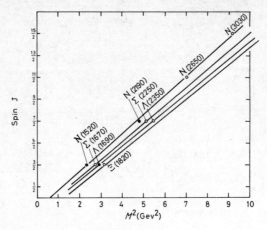

Figure 7.5 Regge trajectories for the odd parity baryon states. ● represents confirmed, and ○ unconfirmed, spin. (N.B. The spin of Σ(1670) is now confirmed as 3/2)

Figure 5. These are called *Regge trajectories,* and the plot is called a *Chew-Frautschi diagram*. In 1959 Regge performed a pioneer analysis of the Schrödinger equation in which he treated angular momentum as a *complex variable*. The implications of this were developed by Chew, Frautschi and others who constructed a viable theory of strong interaction dynamics with spin appearing as a complex variable. Any hadron, on this philosophy, is a pole in the scattering amplitude corresponding to a physical (integral or half integral) value of the spin — a so-called *Regge pole*. A Regge trajectory is simply a manifestation of the dependence of the spin J on energy, or equivalently, on mass; $J(m^2)$. As we move along a trajectory, whenever J passes through integral (or half-odd integral) values, a hadron appears.† The so-called "stable" hadrons are simply the lowest lying members on the trajectory. They are not fundamentally different from any other members.

4) The *lack of exotic states* tells against a picture of resonances as being composite. Why should there be a composite state of π^+ and π^-, but not one of π^+ and π^+, when π^+ and π^- are so intimately related? Pictures like Figures 2 and 4, reminiscent as they are of atomic and nuclear spectroscopy, are far more suggestive of a model of *excitation* than of compositeness for the hadrons.

For these reasons, many physicists believe in what Chew has called a "nuclear democracy": no one hadron is more elementary than any other. The only difference between resonances and stable particles is that resonances are heavier.

---

† Actually, the hadrons appear at intervals $\Delta J = 2$ on any given trajectory.

FURTHER READING

Chew, G. F., M. Gell-Mann and A. H. Rosenfeld, "Strongly Interacting Particles", *Scientific American*, February 1964.
Jacob, M., Contribution to M. Jacob and G. F. Chew, *Strong Interaction Physics*, Benjamin, 1964.
Miller, D. H., "The Elementary Particles with Strong Interactions. I The Baryon Systems, II The Meson Systems." in *High Energy Physics*, vol. IV (E. H. S. Burhop, ed.), Academic Press, 1969.
Omnès, R., *Introduction to Particle Physics*, Wiley-Interscience 1971, Chapters 8, 9 and 14.
Perkins, D. H., *Introduction to High Energy Physics*, Addison-Wesley 1972, Chapter 7.

CHAPTER 8

Weak Interactions and Parity Violation

8.1 THE THETA-TAU PUZZLE

IN THE YEARS leading up to 1956, two particles, then called θ and τ, were observed with the following decay modes

$$\theta^+ \to \pi^+ + \pi^0$$

$$\tau^+ \to 2\pi^+ + \pi^- \tag{8-1}$$

The θ and τ masses were quoted as $(966.7 \pm 2.0)m_e$ and $(966.3 \pm 2.0)m_e$ and their lifetimes as $(1.21 \pm 0.02) \times 10^{-8}$ sec and $(1.19 \pm 0.05) \times 10^{-8}$ sec. The obvious conclusion was that these were the same particle, decaying in different ways. It was shown by Dalitz, however, that θ and τ have opposite parities, so cannot be the same particle. The reasoning ran as follows.

Consider the decay $\theta^+ \to \pi^+ + \pi^0$. If θ^+ has spin s, then the orbital angular momentum of the pions must be s (since they have no spin), and the final state will have parity $(-1)^2(-1)^s = (-1)^s$. If parity is conserved, this will also be the parity of the initial state θ^+, which will therefore have the possible spin-parity assignments $0^+, 1^-, 2^+, \ldots$. In addition to $\theta^+ \to \pi^+ + \pi^0$, a decay $\theta^0 \to \pi^0 + \pi^0$ was also seen, with θ^0 having a similar mass to θ^+. In this neutral decay, however, the final state of two identical bosons must, by the Pauli principle, have l even, so the spin-parity of θ^0 is restricted to the values $0^+, 2^+, 4^+, \ldots$. The same is also true of θ^+ if θ^+ and θ^0 belong to the same isospin family. θ is then *a positive parity meson with even spin*.

Now consider the decay $\tau^+ \to 2\pi^+ + \pi^-$. To deduce the spin and parity of τ^+, Dalitz devised what is now known as the Dalitz plot, which has been discussed already in connection with other particles (see Section 7.7). Let the spin of τ be J; this is then the total angular momentum of the three pion state. Let the orbital angular momentum of the two π^+ about their centre of mass be l_+, and let l_- be the angular momentum of π^- about the centre of mass of $2\pi^+$, as shown in Figure 1. Then, from the vector addition rule

$$|l_+ - l_-| \leqslant J \leqslant l_+ + l_-.$$

132

WEAK INTERACTIONS AND PARITY VIOLATION 133

Figure 8.1 The relative orbital angular momenta l_+ and l_- in the decay $\tau^+ \to \pi^+\pi^+\pi^-$

Because the two π^+ are identical, l_+ must be even. The possible values of l_+, l_- and J appear in Table 1. The parity P of τ^+ is calculated as follows. The parity of $(2\pi^+)$ is $(-1)^2(-1)^{l_+} = (-1)^{l_+}$. The parity of π^- is -1, so the overall parity is

$$P = (-1)^{l_+}(-1)(-1)^{l_-}$$

$$= (-1)^{l_+ + l_- + 1}.$$

Now if $J \neq 0$, then either l_+ or l_- (or both) $\neq 0$. If, for instance, $l_+ \geq 2$, then there should be no decays when the two are at rest relative to one another (since then $l_+ = 0$). This situation arises when π^- has the maximum possible energy, that is, in the region at the top of the Dalitz plot. Thus, if $l_+ \geq 2$, there should be a depletion of events in this region. Similarly, if $l_- \geq 1$, there should be a depletion of points in the region where π^- is at rest relative to $2\pi^+$, i.e. at the bottom of the Dalitz plot. The experimental plot is shown in Figure 2, and to a high degree of accuracy, the distribution of points is *uniform*. This implies that $l_+ = l_- = 0$, and therefore that τ^+ has $J^P = 0^-$.

θ and τ therefore cannot be the same particle, since $J^P(\tau) = 0^-$, which is not a possible spin-parity for θ. The theta-tau puzzle is then: *are θ and τ the same particle or not?* The evidence both ways seems compelling; they have the same mass and lifetime on the one hand, but on the other hand they have different parities. There are two possibilities:

A) θ and τ are different particles since their parities are different. They *accidently* have the same mass and lifetime. This is called a "parity doublet" theory.

Table 8.1

l_-	l_+	J^P
0	0	0^-
1	0	1^+
2	0	2^-
0	2	2^-
1	2	$1^+, 2^+, 3^+$
2	2	$0^-, 1^-, 2^-, 3^-, 4^-$

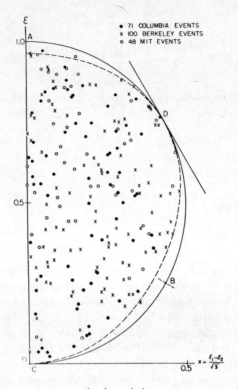

Figure 8.2 Dalitz plot for the decay $\tau^+(K^+) \to \pi^+\pi^+\pi^-$. ϵ is the π^- kinetic energy divided by its maximum possible value, and ϵ_1 and ϵ_2 are the corresponding quantities for the π^+ energies, where $\epsilon_1 > \epsilon_2$. The data points should lie inside the relativistic boundary (dashed curve). The semicircular boundary ABC is due to momentum conservation assuming non-relativistic energies (From J. Orear *et al, Physical Review* **102**, 1676 (1956).)

B) θ and τ are the same particle with a unique mass, lifetime, spin *and parity*, but in one of the decays (1), the parity is *changed*, i.e. is *not conserved*.

Both possibilities involve drastic ideas, but Lee and Yang were bold enough to suggest the more drastic possibility (B), pointing out that to date there existed no evidence that *weak* interactions conserved parity — from the lifetimes of the decays (1), it is clear that they are weak interactions. They then suggested a definitive experiment to find out whether nuclear beta decay conserved parity. The experiment was performed by Wu and collaborators, and spectacularly confirmed that parity was indeed violated in the beta decay of Co^{60}. This vindicated (B), and suggested that θ and τ *are* the same particle, which is now called the K meson, $J^P = 0^-$. The decays (1) are

$$K^+ \to \pi^+ + \pi^0 \tag{8-2}$$

WEAK INTERACTIONS AND PARITY VIOLATION   135

$$K^+ \to \pi^+ + \pi^+ + \pi^- \tag{8-3}$$

so that (2) violates parity, and (3) conserves it. We shall now consider the proposition of Lee and Yang, and the historic experiment of Wu.

8.2 THE DIRECT OBSERVATION OF PARITY VIOLATION

If (B) above is the correct resolution of the θ-τ puzzle, then observation of the decays (1) is itself evidence that parity is not conserved. But it does not correspond to our intuitive notions of what parity nonconservation should look like. We would expect evidence of a *right/left asymmetry*, whereas the decays are evidence of *violation of the parity quantum number*, and require the apparatus of the quantum theory for their interpretation in terms of parity violation. The merit of Lee and Yang's proposal was that it referred to the possibility of the *direct* observation of a right/left asymmetry, independent of the quantum (or any other) theory.

To see that the parity operation changes left into right, consider the right-handed coordinate system in Figure 3(a). Recall that the space inversion or parity operator P is defined by (cf. section 2.6)

$$\begin{pmatrix} x \\ y \\ z \\ t \end{pmatrix} \xrightarrow{P} \begin{pmatrix} -x \\ -y \\ -z \\ t \end{pmatrix} \tag{8-4}$$

or, more succinctly,

$$\mathbf{r} \xrightarrow{P} -\mathbf{r}, \quad t \xrightarrow{P} t. \tag{8-5}$$

The resulting coordinate system is drawn in Figure 3(b). This is left-handed, as may be seen by rotating first about the y axis through $180°$, and then about the z axis through $90°$. The final position, (d), is clearly left-handed. (It is, of course, axiomatic that rotations do not change the handedness of the coordinate system.) By comparing Figures 3(a) and (c), we see that *parity is equivalent to reflection in a plane (mirror)* — in this case the xz plane, which transforms $y \to -y$. In fact we may write

$$P = S_{xz} R_y(\pi) \tag{8-6}$$

where

S_{xz} = reflection in xz plane

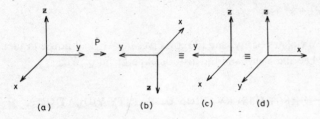

Figure 8.3 The parity operator P transforms the right-handed coordinate system (a) into (b), which by successive rotations about the y axis through π, and the z axis through $\pi/2$, is transformed into (d), which is a left-handed coordinate system

$R_y(\pi)$ = rotation about y axis through π.

P is equivalent to S because R does not change handedness. Because of this, we may represent the parity of a situation by reflecting in a mirror.[†]

To say that a reaction conserves parity is to say that it looks the same in right-handed and left-handed coordinate systems; in other words, that right and left are *indistinguishable*. If parity is not conserved in the reaction, then it looks different in *rh* and *lh* coordinate systems, and we say that right and left are *distinguishable*. As in Chapter 3 we make the association

symmetry → indistinguishability

breaking of symmetry → distinguishability

This can be stated more precisely. What we are concerned with is *observables*, for instance position **r**, momentum **p**, angular momentum **l**, etc. The behaviour of such observables under parity is given in Table 2.2 (p.29). Observables of the form **p.r** do not change sign under P, but those of the form **p.l** or **p.**σ change sign under P. Observation of these latter observables, then, is evidence for parity violation. It is on this fact that the beta decay experiment of Wu et al relies.

8.3 PARITY VIOLATION IN BETA DECAY

We come now to the crucial experiment performed by Wu, Ambler, Hayward, Hoppes and Hudson on the beta decay of Co^{60}

[†] This is true in 3 dimensional space. In 2 dimensional space parity is defined by $\binom{x}{y} \to \binom{-x}{-y}$ which is merely rotation in the plane through π. So *rotation invariance implies parity conservation in* 2 dimensions. The same is true in a 4 dimensional, and in fact in any even dimensional space.

WEAK INTERACTIONS AND PARITY VIOLATION

$$Co^{60} \to Ni^{60} + e^- + \bar{\nu}_e$$

which is of course equivalent to the neutron beta decay

$$n \to p + e^- + \bar{\nu}_e.$$

What was measured was the angular distribution of electrons emitted from *polarised* Co^{60} nuclei. The nuclei of a cobalt salt were polarised by a magnetic field, and the sample was held at a very low temperature to reduce thermal motion. Part of the ingenuity of the experimenters was the way in which they succeeded in simultaneously measuring the electron distribution and lowering the temperature of the salt to the required $0.01°K$. Compared with the complexity of the of the experimental arrangement, the results they got were extremely straightforward; they observed a *preferential direction of emission* of the electrons. This is illustrated in Figure 4. Without reference to any theory of beta decay, it is obvious that this decay violates parity, for in the mirror reflection in Figure 4, the current in the solenoid is reversed, thus reversing the magnetic field and the direction of spin-alignment of the cobalt nuclei. The direction of electron emission is not reversed, however, so that whereas the experiment showed that the electrons were preferentially emitted *antiparallel* to the nuclear spin, the mirror image experiment would show emission *parallel* to the nuclear spin. The experiment and its mirror image are distinguishable, showing that parity is violated.

If σ denotes the nuclear spin, **p** the electron momentum, and $\langle \rangle$ denotes an average value, then the experiment showed that

$$\langle \sigma \cdot \mathbf{p} \rangle < 0.$$

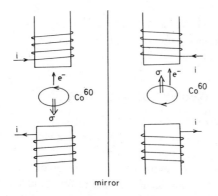

Figure 8.4 Preferential direction of e^- emission in the decay $Co^{60} \to Ni^{60} + e^- + \bar{\nu}$

Under mirror reflection S we have

$$\sigma \xrightarrow{S} -\sigma, \quad \mathbf{p} \xrightarrow{S} \mathbf{p},$$

so

$$\langle \sigma \cdot \mathbf{p} \rangle \xrightarrow{S} -\langle \sigma \cdot \mathbf{p} \rangle$$

Equivalently, under P, $\sigma \to \sigma$, $\mathbf{p} \to -\mathbf{p}$, $\langle \sigma \cdot \mathbf{p} \rangle \to -\langle \sigma \cdot \mathbf{p} \rangle$. If parity were conserved in beta decay, we should have $\langle \sigma \cdot \mathbf{p} \rangle = -\langle \sigma \cdot \mathbf{p} \rangle$, $\langle \sigma \cdot \mathbf{p} \rangle = 0$; there would be isotropic emission of electrons.

Actually, the angular distribution of the electrons is given by

$$4\pi I(\theta) = 1 + A \frac{\langle J_z \rangle}{J} \frac{v}{c} \cos\theta$$

where J is the spin of the Co nucleus, and v/c the helicity h of the electron. (Helicity is $h = \sigma \cdot \mathbf{p}/|\mathbf{p}|$, the component of spin along the direction of motion.) The experiment gave

$$A \approx -1, \tag{8-7}$$

indicating that there is *maximum violation of parity*, since $|A|$ must be less than 1. There is some aesthetic advantage to be gained from the knowledge that beta decay violates parity maximally.

This experiment, together with many others performed in 1957 (see sections 6 and 7 below on π, μ and Λ decay) spectacularly verified Lee and Yang's proposal that weak interactions in general do not conserve parity.

As has been mentioned before, the breaking of a symmetry is associated with a *measurability*. In the case of parity violation it is clear that what is measurable is the absolute difference between right and left. It is an interesting exercise to devise a set of instructions for the performance of the Co^{60} experiment, enabling us to inform the inhabitant of a distant stellar system which is his right hand.

8.4 PARITY VIOLATION AND THE NEUTRINO

The question which now naturally arises is, *why* does beta decay violate parity? Is there any particular circumstance in beta decay on which parity violation may be blamed? First of all, let us note that the way in which parity is violated is that the parity transform of the Wu experiment is an experiment *which does not exist*. (This would be the experiment with the result $\langle \sigma \cdot \mathbf{p} \rangle > 0$.) The quantity the experiment measures is $\sigma \cdot \mathbf{p}$, and for individual particles this quantity is related to the *helicity h*, defined above as

WEAK INTERACTIONS AND PARITY VIOLATION

$$h = \frac{\sigma \cdot \mathbf{p}}{|\mathbf{p}|}. \tag{8-8}$$

It is the projection of spin along the momentum direction. For ordinary spin ½ particles, for instance the proton, there are two spin directions, so the wave function has two components. Dirac's relativistic theory of spin ½ particles, however, predicts a four-component wave function; the other two components correspond to the antiparticle states, again with two spin directions. The four states can conveniently be chosen to be eigenstates of helicity with the eigenvalues $h = \pm 1$, though it should be understood that the helicity of a physical proton is $\pm v/c$ where v is its velocity; in other words, the wave function of a physical proton is a linear combination of wave functions with $h = \pm 1$. Calling $h = +1$ and $h = -1$ respectively right-handed (R) and left-handed (L) states, the proton wave function is then written

$$p = \begin{pmatrix} p_L \\ p_R \\ \bar{p}_L \\ \bar{p}_R \end{pmatrix} \tag{8-9}$$

a 4-component "spinor", consisting of a left-handed proton, a right-handed proton, a left-handed antiproton and a right-handed antiproton. Under parity, left and right are interchanged, but not particle and antiparticle

$$p_L \overset{P}{\leftrightarrow} p_R, \quad \bar{p}_L \overset{P}{\leftrightarrow} \bar{p}_R \tag{8-10}$$

Following the Wu experiment, a *two-component neutrino theory* was proposed by Lee and Yang, by Landau and by Salam. According to this theory, the neutrino wave function is written

$$\nu = \begin{pmatrix} \nu_L \\ - \\ - \\ \bar{\nu}_R \end{pmatrix} \tag{8-11}$$

in other words, there is only a left-handed neutrino and a right-handed antineutrino; the other 2 states *do not exist*.

It is obvious that the very existence of such a neutrino violates parity, since under the parity operation

$$\nu_L \overset{P}{\rightarrow} -(\nu_R); \quad \bar{\nu}_R \overset{P}{\rightarrow} -(\bar{\nu}_L) \tag{8-12}$$

the existing states of the neutrino are transformed into states which do not exist[†]. All interactions involving the neutrino must then *of necessity* violate parity.

Of course, the neutrino wave function could have been chosen to have the other two components

$$\nu' = \begin{pmatrix} - \\ \nu_R \\ \overline{\nu}_L \\ - \end{pmatrix} \tag{8-13}$$

Parity would still be violated. It is for experiment to decide between these two alternatives. If the definition is adopted that the particle emitted with the electron in the beta decay of the neutron is the antineutrino, i.e.

$$n \to p + e^- + \overline{\nu} \tag{8-14}$$

then experiment can decide whether this particle is right- or left-handed. The experiment is described below, and shows that the antineutrino in (14) is right-handed, and the neutrino left-handed; as in (11).

In closing this section it is appropriate to make 3 remarks.

1) The neutral emitted particle in (14) above is defined to be the antineutrino in order that the *conservation of lepton number,* discussed in chapter 1, holds, with the assignments

$$L_{e^-} = L_\nu = 1; \quad L_{e^+} = -1;$$

so that if an electron is a lepton, a neutrino is also a lepton, and an antineutrino an antilepton. This is purely a matter of convention. What is not convention is whether the particle emitted with the electron is right- or left-handed.

2) The 2-component neutrino theory and the law of lepton conservation together imply that *the neutrino has zero rest-mass.* This is because if it had non-zero mass it would not travel at the speed of light; in that case it would be possible to *overtake* it, and it would then be observed to change from a left-handed particle to a right-handed one, since its momentum, but not its spin, would change sign. According to the 2-component theory, this means that by overtaking a neutrino (ν_L) we would see it as an antineutrino ($\overline{\nu}_R$), which violates the law of lepton conservation. It must therefore be impossible to overtake the neutrino,

---

[†] Or, equivalently, into states which exist but do not interact. The difference is metaphysical; existence is only apprehended through interaction.

WEAK INTERACTIONS AND PARITY VIOLATION 141

which means that it must travel at the speed of light, and therefore have zero mass.† (The experiments give $m_{\nu_e} < 60$ eV.)

3) This explanation of parity violation in beta decay gives no indication of why the nonleptonic decays $K \to 2\pi$, $\Lambda \to p + \pi^-$, etc, violate parity. They remain a comparative mystery.

8.5 THE HELICITY OF THE NEUTRINO

As promised we shall now consider the experimental evidence that the neutrino is left-handed, i.e. that it has helicity $h = -1$. The experiment was performed by Goldhaber, Grodzins and Sunyar, and in essence is as follows. Instead of (14) consider the equivalent reaction

$$e^- + p \to n + \nu$$

This is what happens in the capture of a K-shell electron by a nucleus A

$$e^- + A \to B^* + \nu,$$
$$\quad\; J{=}0 \quad\; J{=}1$$

leaving an excited nucleus B^*. Suppose that these nuclei have spins 0 and 1, as

---

† The Dirac equation may be written (where $c = 1$)

$$(E - \sigma.\mathbf{p})u = mv$$

$$(E + \sigma.\mathbf{p})v = mu$$

where u and v are both two component wave functions, and σ are the 2×2 Pauli matrices. For zero mass the above equations *decouple* to give

$$(E - \sigma.\mathbf{p})u = 0$$

$$(E + \sigma.\mathbf{p})v = 0.$$

These are called the Weyl equations. The eigenvalue equation for them becomes $E^2 = \mathbf{p}^2$ or $E = \pm|\mathbf{p}|$ (since $\sigma^2 = 1$). This is as expected for zero mass. The solution $E = -|\mathbf{p}|$ is interpreted to correspond to antiparticles. Following the definition (8) of helicity we have

$$hu = \frac{\sigma.\mathbf{p}}{|\mathbf{p}|}u = \frac{E}{|\mathbf{p}|}u = \begin{cases} +u \text{ for particles} \\ -u \text{ for antiparticles.} \end{cases}$$

The opposite is true for the ν solution. u and v correspond to the two choices (13) and (11) respectively for a two component massless spin ½ particle. Weyl discovered these equations long before parity violation was discovered, and they were disregarded for that same reason.

marked, and that B^* then decays to its spin-zero ground state by photon emission

$$B^* \to B + \gamma.$$

The overall reaction is $e^- + A \to B + \nu + \gamma$ through a *resonant* channel B^*. It can be shown that the resonance condition is fulfilled when the photon and neutrino move away in *opposite directions*.

Now consider the conservation of J_z, and note that although the photon has spin 1, the fact that it is massless means that $J_z = \pm 1$; the $J_z = 0$ state does not exist. Take the z axis to be the direction of neutrino emission. If the neutrino is left-handed, it has $J_z = -\tfrac{1}{2}$ and conservation of J_z demands the following assignments (see Figure 5(a))

$$e^- + A \to B + \nu_L + \gamma$$
$$J_z: \tfrac{1}{2} \quad 0 \quad 0 \quad -\tfrac{1}{2} \quad 1$$

so that γ has $J_z = 1$ but is moving along the negative z axis, and so has negative helicity — like the neutrino. On the other hand, if the neutrino is right-handed ($h = 1$) we must have (Figure 5(b))

$$e^- + A \to B + \nu_R + \gamma$$
$$J_z: -\tfrac{1}{2} \quad 0 \quad 0 \quad \tfrac{1}{2} \quad -1$$

and the photon has positive helicity, again like the neutrino. In either case, then, measuring the photon helicity is tantamount to measuring the neutrino helicity.

Goldhaber *et al* rather cleverly found nuclei with the required spins: Eu^{152} has $J^P = 0^-$, and gives an excited 1^- state Sm^{152*} on K capture; this decays to the 1^+ ground state. The emitted photon was found to have negative helicity, so it follows that the neutrino also has negative helicity, i.e. is left-handed.

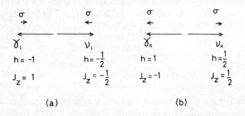

(a)   (b)

Figure 8.5 Emission of ν and γ in the experiment to determine the neutrino helicity

WEAK INTERACTIONS AND PARITY VIOLATION 143

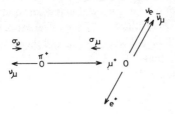

Figure 8.6 π-μ-e decay sequence

8.6 π-μ-e DECAY SEQUENCE

As mentioned above, nuclear beta decay was not the only decay observed to violate parity. Lee and Yang's suggestion was a general one, that weak interactions should be expected to violate parity. Among the experiments performed was one on the decay sequence

$$\pi^+ \to \mu^+ + \nu_\mu$$

$$\mu^+ \to e^+ + \nu_e + \bar{\nu}_\mu.$$

If parity is not conserved in π decay, $\sigma_\mu \cdot \mathbf{p}_\mu$ could be non-zero, i.e. the muon could be longitudinally polarised. In fact, if ν_μ is left handed, μ^+ will also be left-handed, to conserve angular momentum − see Figure 6.

The muons will slow down and stop before they decay, but in a suitable material their spin direction will be unchanged, so we have a source of polarised muons at rest. If in the decay of μ^+ parity is also not conserved, there will be a non-zero value of $\langle \sigma_\mu \cdot \mathbf{p}_e \rangle$, i.e. a forward-backwards asymmetry in the positron distribution with respect to the original μ^+ momentum. Friedman and Telegdi, and Garwin, Lederman and Weinrich observed such an asymmetry only a few days after the results on beta decay were known. These results showed that parity is violated in *both* the decays $\pi \to \mu\nu$ and $\mu \to e\bar{\nu}\nu$.

8.7 Λ NONLEPTONIC DECAY

The dominant decay mode of Λ,

$$\Lambda \to p + \pi^-,$$

provides another example, besides $K^+ \to 3\pi$, of a *nonleptonic* decay which may be tested for parity violation. It has in fact been shown to violate parity, using the following reasoning.

ELEMENTARY PARTICLES AND SYMMETRIES

Λ and p both have spin ½, so the orbital angular momentum in the final state may be $l = 0$ or $l = 1$. These states will have different parity, but if the decay violates parity, both will be allowed. Let us call them s-wave and p-wave states. Let us arbitrarily assume that Λ has J_z = ½; the final state then must also have J_z = ½. If the final state has $l = 0$ its wave function is

$$\psi_s = a_s Y_0^0 \psi(½, ½) \tag{8-15}$$

where

a_s = amplitude for s-wave decay

Y_0^0 = orbital wave function for $l = 0, m = 0$

$\psi(½, ½)$ = proton spin wave function for $s = ½, m = ½$.

If $l = 1$, then the p-wave final state will be

$$\psi_p = a_p [c_1 Y_1^1 \psi(½, -½) + c_2 Y_0^1 \psi(½, ½)]$$

where the symbols have a parallel meaning to those in (15). c_1 and c_2 are Clebsch-Gordon coefficients, which on comparison with equation (3A-25) are $c_1 = \sqrt{2/3}$, $c_2 = -\sqrt{1/3}$. Y_m^l are the spherical harmonics, and have the values

$$Y_0^0 = \frac{1}{\sqrt{4\pi}}, \quad Y_0^1 = \sqrt{\frac{3}{4\pi}} \cos\theta, \quad Y_1^1 = -\sqrt{\frac{3}{8\pi}} \sin\theta e^{i\varphi},$$

so that

$$\psi_s = \frac{a_s}{\sqrt{4\pi}} \psi(½, ½), \quad \psi_p = \frac{-a_p}{\sqrt{4\pi}} [\sin\theta e^{i\varphi} \psi(½, -½) + \cos\theta \psi(½, ½)].$$

The final wave function is

$$\psi = \psi_s + \psi_p = \frac{1}{\sqrt{4\pi}} [(a_s - a_p \cos\theta) \psi(½, ½) - a_p \sin\theta e^{i\varphi} \psi(½, -½)],$$

and the intensity distribution is

$$\psi^* \psi \propto |a_s - a_p \cos\theta|^2 + |a_p|^2 \sin^2\theta = |a_s|^2 + |a_p|^2 - 2 Re(a_s a_p^*) \cos\theta$$

WEAK INTERACTIONS AND PARITY VIOLATION

which follows because $\psi(\tfrac{1}{2}, \tfrac{1}{2})$ and $\psi(\tfrac{1}{2}, -\tfrac{1}{2})$ are orthogonal. This expression is of the form $1 + a\cos\theta$ where

$$a = -\frac{2Re(a_s a_p^*)}{|a_s|^2 + |a_p|^2} \tag{8-16}$$

is the asymmetry, or up/down, parameter for the decay. θ is the angle between π^- and the Λ spin. Since in the actual case Λ is not completely polarised, the angular distribution will be

$$1 + aP\cos\theta$$

where P is the transverse polarisation of Λ. θ is now effectively the angle between the direction of the final π^- and the production plane of Λ, as shown in Figure 7. The production plane is defined by the direction of Λ and the initial pion in the production reaction

$$\pi^- + p \to \Lambda + K^0.$$

It is clear that the product of three momenta changes sign under parity, as it is clear that, referring to Figure 7, if there are more "up" decays than "down" decays, the decays will look different in the mirror, i.e. parity is violated. Early experiments on Λ decay by Crawford *et al*, Eisler *et al*, and Adair and Leipuner, gave $aP \approx +0.7$, confirming parity violation.

A note on the spins of Λ, Σ and Ξ

Lee and Yang made use of this experiment, and proved that if the spin of Λ in the above decay is s, then

$$|a\overline{P}| \leq \frac{1}{2s} \tag{8-17}$$

where $\overline{P}$ is the mean value of P. For $s = \tfrac{1}{2}$, (17) reduces to $|a\overline{P}| \leq 1$, and for

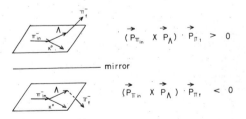

Figure 8.7 Up-down asymmetry in Λ production and decay

$s = \frac{3}{2}$ to $|a\bar{P}| \leq \frac{1}{3}$. The experimental value $a\bar{P} \sim 0.7$ above clearly indicates that Λ has spin ½. In addition, experiments have shown that

$$a\bar{P}(\Sigma^+ \to p + \pi^0) = 0.75 \pm 0.17$$

$$P(\Xi^- \to \Lambda + \pi^-) = 0.52 \pm 0.26$$

indicating that Σ^+ has spin ½, and that Ξ^- probably also has.

This is not the only method of determining the spins of Λ, Σ and Ξ, but it has the advantage over other methods that the theorem of Lee and Yang is independent of the spins of π and K, which, in other determinations, have to be assumed to be zero. In fact, the spins of Λ, Σ and Ξ are all believed to be ½.

8.8 CP CONSERVATION IN BETA DECAY

The charge conjugation operator C was defined in section 2.13. It changes particle ↔ antiparticle without changing the helicity, so on the p spinor (9) it has the following effect

$$p_L \stackrel{C}{\leftrightarrow} \bar{p}_L; \quad p_R \stackrel{C}{\leftrightarrow} \bar{p}_R.$$

Consequently the neutrino wave function (11) is transformed into

$$\nu_L \stackrel{C}{\to} -(\bar{\nu}_L); \quad \bar{\nu}_R \stackrel{C}{\to} -(\nu_R)$$

into a non-existent state. We therefore expect that *beta decay violates C*. Just as it provides a means of distinguishing between left and right, it also provides a means of distinguishing between particles and antiparticles.

EXERCISE   Verify directly that changing particles to antiparticles in the Co^{60} experiment changes the sign of $\langle \sigma \cdot \mathbf{p} \rangle$.

Combining the operations C and P, however, gives

$$\nu_L \stackrel{CP}{\leftrightarrow} \nu_R$$

both of which exist. This suggests that beta decay conserves CP, and in fact the present theory of the decay, in full agreement with experiment, is CP conserving. That is to say, on changing left ↔ right *and* particles ↔ antiparticles, the Wu experiment will look identical

$$\langle \sigma \cdot \mathbf{p} \rangle \stackrel{CP}{\to} \langle \sigma \cdot \mathbf{p} \rangle;$$

WEAK INTERACTIONS AND PARITY VIOLATION

the preferential direction of electron emission is backward in both cases.

One consequence of this is that the Co^{60} experiment cannot be used to define left and right, *and* particle and antiparticle simultaneously. A convention must be fixed for one of them before the other can be defined. Thus in communicating to the inhabitant of a distant galaxy over the Co^{60} experiment, if he happens to be made of antimatter relative to us, what we define to be right, he will get to be left.

This chapter started by discussing the possibility of parity violation in θ and τ decays (K decays). It was then shown that beta decay (as well as other decays) violates parity; but conserves CP. It is, then, logical to return to the K decays and enquire whether they also conserve CP. This requires the development of a theory of K mesons which is the subject of the next chapter.

FURTHER READING

Lee, T. D. Nobel prize lecture 1957, in *Nobel Lectures; Physics: 1942–1962*, Elsevier, 1964.

Lee, T. D. and C. N. Yang, "Parity Nonconservation and a Two-component Theory of the Neutrino", *Physical Review* **105**, 1671 (1957).

Lee, T. D. and C. N. Yang, "Question of Parity Conservation in Weak Interactions", *Physical Review* **104**, 254 (1956).

Wu, C. S., E. Ambler, R. Hayward, D. Hoppes and R. Hudson, "Experimental Test of Parity Conservation in Beta Decay", *Physical Review,* **105**, 1413 (1957).

Wu, C. S., "Beta Decay", in *Weak Interactions and High Energy Neutrino Physics,* Proceedings of the International School of Physics "Enrico Fermi", course 32, Academic Press, 1966.

CHAPTER 9

K Meson Decays and CP Violation

AS A RESULT of the discovery of parity violation, physicists were forced to take a hard look at the other discrete symmetries, wondering which, if any, of them were to remain inviolate. We have already seen that the two component neutrino theory implies that C, as well as P, is violated, but that CP is conserved. But what about T? (We recall from Chapter 2 that because T changes a wave function into its complex conjugate there is no conserved quantum number associated with T invariance. For this reason, T invariance is a bit more difficult to test experimentally than P or C invariance.) One of the most remarkable off-shoots of these investigations was the so-called CPT theorem. This was proved by Schwinger, Lüders and Pauli, and plays a central part in any discussion of discrete symmetries. The present chapter begins with a description of the CPT theorem, and continues with the theory of K decays proposed by Gell-Mann and Pais. This theory enables us to determine whether K decays, like beta decay, conserve CP. It turns out that most K decays do conserve CP, but that some rare ones do not. The chapter finishes with an outline of the so far unsolved problem of CP violation in K decays.

9.1 CPT THEOREM

As its name implies, the CPT theorem concerns the joint product transformation CPT, changing $\mathbf{r} \to -\mathbf{r}$, $t \to -t$, and particles ↔ antiparticles. The theorem states that *any* quantum theory of fields which is compatible with special relativity and assumes only local interactions is *automatically* invariant under CPT.

All interactions, therefore, even the weak ones, are invariant under CPT. The implication of the theorem is clearly that the product CPT is more fundamental than any of the operators C, P and T separately. Experimental evidence for the breakdown of CPT invariance would be much more far-reaching in its implications than evidence of non-invariance under C, P and T separately (disturbing though they may seem!), since it would strike at the very fundamentals of quantum field theory.

One consequence of the CPT theorem is that if an interaction is invariant under one of the transformations C, P or T, it must be invariant under the product of the other two; and, similarly, that if it is non-invariant under one of them, it must be non-invariant under the product of the other two. Thus, the fact that beta decay violates P means that it violates CT; and that it is invariant under CP implies that it is invariant under T: it does not distinguish between time running forwards and time running backwards. Invariance of beta decays under T implies that the Fermi and Gamow-Teller coupling constants (see section 10.2) should be 180° out of phase. The experimental phase difference is $180° - (1.3 \pm 1.3)°$. The decay $\Lambda \to p + \pi^-$ also appears to be invariant under T; the condition for T invariance is that the s-wave and p-wave amplitudes are in phase, and the measured phase difference is $2.8 \pm 4°$. We shall discuss the question of CP (and by implication T) invariance of K decays below.

Another important consequence of the CPT theorem is that particles and antiparticles have equal masses and lifetimes. As may be expected, this would also be a consequence simply of C invariance, if that were exact, but the fact that it also follows from CPT invariance means that the breakdown of C invariance (for instance in beta decay) does *not* imply the breakdown of this equality, which provides one of the best tests of the CPT theorem. The experimental evidence may be summarised as follows.

a) e^+ and e^- have equal masses to within 1 part in 10^5 — this is a test of CPT for electromagnetic interactions. To support the CPT invariance of strong interactions there is the fact that K^+ and K^-, and π^+ and π^-, have equal masses to within 1 part in 10^3. But the best test of CPT is a measure of the mass difference between K^0 and $\bar{K}^0$. This cannot be measured directly, but it is necessarily less than the mass difference between K_L^0 and K_S^0 (see below), which gives

$$\frac{m(K^0) - m(\bar{K}^0)}{m(K)} \leqslant \frac{m(K_L^0) - m(K_S^0)}{m(K)} = 0.71 \times 10^{-14}.$$

This is a very stringent limit, and provides evidence for the CPT invariance of strong, electromagnetic and weak interactions, to a high degree of accuracy.

b) The lifetimes of the particles and antiparticles $\mu^\pm$, $\pi^\pm$ and $K^\pm$ are known to be equal to within 5 parts in 10^4.

c) CPT invariance also implies that e^+ and e^-, and μ^+ and μ^-, have equal and opposite magnetic moments. The measured differences are

$|\mu(\mu^+)| - |\mu(\mu^-)| = (-1.2 \pm 1.6) \times 10^{-6};$

$|\mu(e^+)| - |\mu(e^-)| = (1.6 \pm 2.2) \times 10^5.$

In conclusion, there is no known violation of CPT invariance.

9.2 GELL-MAN–PAIS THEORY OF NEUTRAL KAONS

Let us now return to K decays. We have learned already that θ decay violates, and τ decay conserves P. In addition, we saw in the last chapter that

$$\theta^+ = \tau^+ = K^+, \tag{9-1}$$

the θ and τ decays merely being the 2π and 3π modes of K^+ decay. The question arises, what are θ^0 and τ^0

$$\left.\begin{matrix}\theta^0\\ \tau^0\end{matrix}\right\} = \left.\begin{matrix}K^0\\ \overline{K^0}\end{matrix}\right\} ? \tag{9-2}$$

The $|\Delta S| = 1$ rule allows both K^0 and $\overline{K^0}$ to decay into pions, so there appears to be some freedom about what to choose for θ^0 and τ^0. This question was taken up by Gell-Mann and Pais in 1955, before it was realised that θ decay violates parity (and charge parity – see below). Actually, Gell-Mann and Pais assumed that K decays conserve C, but this is not essential to their main argument, which runs as follows.

Consider the particles K^0 and $\overline{K^0}$. Since they have $S = 1$ and $S = -1$ respectively, one is the antiparticle of the other. For comparison, consider the neutron and antineutron, n and $\bar{n}$. They are also electrically neutral; they have $B = 1$ and $B = -1$, and one is the antiparticle of the other. There is a crucial difference, however, between the two cases; B is an *absolutely conserved* quantum number, whereas S is only conserved by strong and electromagnetic interactions. The weak interactions induce the $|\Delta S| = 1$ decays

$$K^0 \to 2\pi, \quad \overline{K^0} \to 2\pi.$$

In addition, the reverse of the second process must also be possible,

$$2\pi \to \overline{K^0},$$

at least as a virtual process (since it is endothermic), and combined with the first decay it gives the overall transition

$$K^0 \to 2\pi \to \overline{K^0}$$
$$\uparrow$$
virtual

i.e., $K^0 \leftrightarrow \overline{K^0}$; the particle and antiparticle *mix* through the weak interactions. In contrast, the neutron and antineutron never mix, because they have different values of B, and there is no interaction which changes B:

K MESON DECAYS AND CP VIOLATION

$$K^0 \leftrightarrow \overline{K^0} \quad ; \quad n \not\leftrightarrow \bar{n}$$
$$S=1 \quad S=-1 \quad B=1 \quad B=-1$$

weak interactions

This observation forms the cornerstone of the Gell-Mann–Pais theory.
Now consider the effect of this mixing on the decay into pions. Let the amplitude for $K^0 \to 2\pi$ be A. It then follows from the *CPT* theorem that the amplitude for $\overline{K^0} \to 2\pi$ is $-A$. By simple addition and subtraction, the mixtures in Table 1 have amplitudes for decay into 2π of $\sqrt{2}A$ and 0; in other words, there are two mixtures, one of which does, and the other of which does not, decay into 2π. (The addition of amplitudes A rather than of probabilities $|A|^2$ is, as the reader will recall, the basic postulate of the quantum theory. It is known as the *superposition principle,* and accounts for the fact that particles may behave like waves.) So, defining

$$K_1^0 = \sqrt{\tfrac{1}{2}}(K^0 - \overline{K^0}) \tag{9-3}$$

$$K_2^0 = \sqrt{\tfrac{1}{2}}(K^0 + \overline{K^0}) \tag{9-4}$$

we see that

$$K_1^0 \to 2\pi, \quad K_2^0 \not\to 2\pi. \tag{9-5}$$

Since 2π is the most easily available decay mode of kaons, (5) implies that K_1^0 is short-lived, but K_2^0 is long-lived. K_2^0 will decay into other modes, notably 3π, but this takes longer because of the smaller phase space;

$$\tau_{K_1^0} \ll \tau_{K_2^0}.$$

This prediction of *two distinct lifetimes for the decays of kaons* was the principal result obtained by Gell-Mann and Pais. Equations (3) and (4) may be inverted to give

Table 9.1

Decay	Amplitude
$K^0 \to 2\pi$	A
$\overline{K^0} \to 2\pi$	$-A$
$\sqrt{\tfrac{1}{2}}(K^0 - \overline{K^0}) \to 2\pi$	$\sqrt{2}A$
$\sqrt{\tfrac{1}{2}}(K^0 + \overline{K^0}) \to 2\pi$	0

$$K^0 = \sqrt{\tfrac{1}{2}}(K_1^0 + K_2^0) \tag{9-6}$$

$$\overline{K^0} = \sqrt{\tfrac{1}{2}}(K_2^0 - K_1^0). \tag{9-7}$$

From (3) and (4), K_1^0 and K_2^0 are distinct states, but they are *not* particle and antiparticle. They have no well-defined value of S. This does not matter, however, because weak interactions do not conserve S, so states which decay via weak interactions do not need to be eigenstates of S. The important point is that K_1^0 and K_2^0 are the states which decay (by the weak interactions), whereas K^0 and $\overline{K^0}$ are the states which are produced (by the strong interactions). *The (decaying) states with definite lifetimes are not the (produced) states with definite strangeness, but linear mixtures of them.*

The consequences of this are very interesting, not to say exotic. Consider the production of a K^0 by $\pi^- p$ scattering

$$\pi^- + p \rightarrow \Lambda + K^0$$
$$ \downarrow \sqrt{\tfrac{1}{2}}(K_1^0 + K_2^0)$$

where we have used (6) to express K^0 in terms of K_1^0 and K_2^0. The K_1^0 component now decays out of the beam into 2π, as shown in Figure 1. A long way from the source, most of the K_1^0 component will have decayed away and the beam will be more or less pure K_2^0, which has a much longer lifetime. From (4), this means the beam will be an *equal mixture of K^0 and $\overline{K^0}$*, whereas when it was produced it was pure K^0. K^0 has been partially *converted* into $\overline{K^0}$. This prediction may serve as a decisive test for the theory, since $\overline{K^0}$ may be detected by the reaction

$$\overline{K^0} + p \rightarrow Y_1^* + \pi \tag{9-8}$$
$$\phantom{\overline{K^0} + p \rightarrow } \downarrow \begin{cases} \Sigma + \pi \\ \Lambda + \pi \end{cases}$$

which will occur if a second target is placed in the beam. Note that Y_1^* cannot be produced from a K^0 since the strangeness is wrong, so the detection of $(\Sigma + \pi)$ or $(\Lambda + \pi)$ is proof that there is $\overline{K^0}$ in the beam. This test, proposed by Pais and Piccioni, has been proved positive by the detection of the final decay products in the Berkeley liquid hydrogen bubble chamber.

After passing through the second target, which from (8) absorbs $\overline{K^0}$, the beam is now pure K^0, which from (6) is an equal mixture of K_1^0 and K_2^0; in other words K_1^0 has been *regenerated* from a pure K_2^0 beam. As before, the K_1^0 component now starts to decay, so that far from the second target the beam is again pure K_2^0. *Decay in a vacuum converts K^0 into $\overline{K^0}$, and passage through a material medium*

K MESON DECAYS AND CP VIOLATION 153

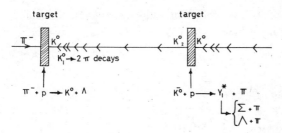

Figure 9.1 Conversion of K^0 to $\overline{K^0}$ and regeneration of K_1^0 from a K_2^0 beam

regenerates K_1^0 from K_2^0. These conversions and regenerations may continue *ad infinitum*, though, of course, the intensity of the beam decreases all the time.

It is worth emphasising that what makes this theory so attractive is that it follows very directly from the quantum theory in the form of the superposition principle. Its experimental verification, therefore, may be counted as one of the most dramatic demonstrations there is of the quantum theory.

9.3 K MESON INTERFEROMETRY AND THE K_1-K_2 MASS DIFFERENCE

Let us consider the process of conversion quantitatively. If K_1^0 has a lifetime τ_1, the associated decay width is $\Gamma_1 = \hbar/\tau_1$ which, in "natural" units where $\hbar = 1$, becomes $\Gamma_1 = 1/\tau_1$, and the wave function for a K_1^0 meson at time t is related to its wave function at $t = 0$ by

$$\psi(K_1, t) \propto \psi(K_1, 0) e^{-\frac{1}{2}\Gamma_1 t}.$$

The ½ occurs so that the probability P of finding a K_1 meson after time t is $P = |\psi(K_1, t)|^2 \propto \exp(-\Gamma_1 t)|\psi(K_1, 0)|^2$. If the mass of the meson is m_1 the above equation becomes

$$\psi(K_1, t) = \psi(K_1, 0) e^{(-im_1 - \frac{1}{2}\Gamma_1)t};$$

Table 9.2

	K mesons with strong interactions		K mesons with weak interactions	
	K^0	$\overline{K^0}$	K_1^0	K_2^0
	1	−1	−	−
τ(sec)	−	−	8×10^{-11}	5×10^{-8}

the mass just causes an oscillation in the wave function. Similarly the amplitude for finding a K_2 at time t is

$$\psi(K_2, t) = \psi(K_2, 0)e^{(-im_2 - \frac{1}{2}\Gamma_2)t}.$$

Experimentally

$$\tau_1 \sim 10^{-10} \text{ sec}, \quad \tau_2 \sim 5 \times 10^{-8} \text{ sec},$$

so that

$$\Gamma_1 \sim 10^{10} \text{ sec}^{-1} \sim 6.6 \times 10^{-6} \text{ eV}$$

$$\Gamma_2 \sim 2 \times 10^7 \text{ sec}^{-1} \sim 1.2 \times 10^{-8}$$

$$\Gamma_1 \approx 500\Gamma_2.$$

The above conversion from an inverse time to an energy is made by multiplying by Planck's constant $\hbar = 1.05 \times 10^{-34}$ Js $= 6.58 \times 10^{-16}$ eVs

From (7), the amplitude for finding a $\overline{K^0}$ at time t is

$$\psi(\overline{K^0}, t) = \sqrt{\tfrac{1}{2}}[\psi(K_2, t) - \psi(K_1, t)]$$

$$= \sqrt{\tfrac{1}{2}}[\psi(K_2, 0)e^{(-im_2 - \frac{1}{2}\Gamma_2)t} - \psi(K_1, 0)e^{(-im_1 - \frac{1}{2}\Gamma_1)t}] \quad (9\text{-}9)$$

Now suppose that at $t = 0$ a K^0 is produced by the reaction $\pi^- \to K^0 + \Lambda$, as in Figure 1. K^0 is an equal mixture of K_1 and K_2 so we have

$$\psi(K_1, 0) = \psi(K_2, 0) = \psi_0;$$

so from (9)

$$\psi(\overline{K^0}, t) = \frac{\psi_0}{\sqrt{2}}[e^{(-im_2 - \frac{1}{2}\Gamma_2)t} - e^{(-im_1 - \frac{1}{2}\Gamma_1)t}]$$

and the probability of finding a $\overline{K^0}$ at time t is

$$P(\overline{K^0}, t) = |\psi(\overline{K^0}, t)|^2$$

$$= \frac{|\psi_0|^2}{2}[e^{-\Gamma_2 t} + e^{-\Gamma_1 t} + 2\text{Re}\, e^{(-im_2 + im_1)t} e^{-\frac{1}{2}(\Gamma_1 + \Gamma_2)t}]$$

$$= \frac{|\psi_0|^2}{2} [e^{-\Gamma_2 t} + e^{-\Gamma_1 t} + 2\cos(\Delta mt)e^{-\frac{1}{2}(\Gamma_1+\Gamma_2)t}] \tag{9-10}$$

where $\Delta m = m_2 - m_1$ and the formula $|x+y|^2 = |x|^2 + |y|^2 + 2Re(x^*y)$ has been used. (10) is the sum of two exponential decays and an interference term involving Δm. For times of the order of τ_1 the function in brackets is plotted in Figure 2, and it is seen that the number of $\overline{K^0}$ mesons in the beam depends sensitively on Δm. This number may be determined by absorbing the beam on a target, as in Figure 1, placed at varying distances from the point of production of the K^0. In this way a value for Δm may be found. ((10) only depends on the magnitude, not the sign, of Δm.) There is rather a spread in the experimental values, but they are of the order $|\Delta m| \sim 0.5\ \Gamma_1$. A more complicated interference experiment has been performed by Piccioni and by Good et al, in which not only the magnitude but also the sign of Δm is measured. Lee and Wu give an average value for all the experiments of

$$\Delta m = (0.5 \pm 0.1)\Gamma_1 \approx 10^{-5}\ \text{eV}. \tag{9-11}$$

so that K_2 is heavier than K_1. Finally, let me repeat that there is no need for K_1 and K_2 to have the same mass, since they are not particle and antiparticle. Their different masses are due to their different decay modes.

9.4 K_1-K_2 MASS DIFFERENCE AS EVIDENCE FOR THE $|\Delta S| = 1$ SELECTION RULE

If weak interactions obey $|\Delta S| = 1$ then $K^0 - \overline{K^0}$ mixing occurs as a *second order*

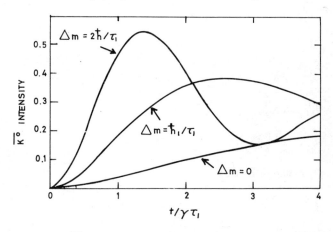

Figure 9.2 $\overline{K^0}$ intensity as a function of flight time for a beam initially K^0

effect, for instance via a 2π intermediate state $K^0 \leftrightarrow 2\pi \leftrightarrow \overline{K^0}$. Each $|\Delta S| = 1$ weak process has strength 0.1 G where $G \sim 10^{-5} m_p^{-2}$ is the weak coupling constant for $\Delta S = 0$ processes. In this case the mass difference will be expected to be of the order

$$|\Delta m| \sim m_K^5 (0.1G)^2 \sim 10^{-4} \text{ eV}$$

where m_K is put in to get the dimensions into those of a mass. This estimate is indeed very close to (11).

On the other hand, if the weak interactions have a $\Delta S = 2$ component, $K^0 \leftrightarrow \overline{K^0}$ mixing may occur as a *first order effect*, and the estimated mass difference will be

$$|\Delta m| \sim m_K^3 (0.1G) \sim 140 \text{ eV};$$

many orders of magnitude greater than the experimental value. This constitutes the best evidence there is that weak interactions do not have a $\Delta S = 2$ component. The argument is due to Okun' and Pontecorvo.

9.5 CONSEQUENCES OF CP INVARIANCE IN K DECAYS

The operator C is defined by

$$CK^0 = \overline{K^0}, \quad C\overline{K^0} = K^0.$$

It follows from (3) and (4) that

$$CK_1^0 = -K_1^0, \quad CK_2^0 = K_2^0. \tag{9-12}$$

Since K^0 and $\overline{K^0}$ both have negative parity

$$PK^0 = -K^0, \quad P\overline{K^0} = -\overline{K^0}$$

then so do K_1^0 and K_2^0

$$PK_1^0 = -K_1^0, \quad PK_2^0 = -K_2^0. \tag{9-13}$$

Combining (12) and (13) gives

$$CPK_1^0 = K_1^0 \tag{9-14}$$

$$CPK_2^0 = -K_2^0. \tag{9-15}$$

K MESON DECAYS AND CP VIOLATION

K_1^0 and K_2^0 are eigenstates of *CP* with the eigenvalues +1 and −1.

Now let us turn to the pions. Consider first a 2π state ($\pi^+\pi^-$) with orbital angular momentum l. We have

$$P(\pi^+\pi^-) = (-1)^2(-1)^l(\pi^+\pi^-) = (-1)^l(\pi^+\pi^-),$$

and, since *C* interchanges π^+ and π^-,

$$C(\pi^+\pi^-) = (\pi^-\pi^+) = (-1)^l(\pi^+\pi^-).$$

It follows that

$$CP(\pi^+\pi^-) = (-1)^{2l}(\pi^+\pi^-) = +(\pi^+\pi^-). \tag{9-16}$$

Now consider a state ($\pi^+\pi^-\pi^0$) where the angular momentum between π^+ and π^- is l, and that between π^0 and the centre of mass of π^+ and π^- is l' (cf. Figure 8-1). We have

$$P(\pi^+\pi^-\pi^0) = (-1)^3(-1)^{l+l'}(\pi^+\pi^-\pi^0)$$

$$= (-1)^{l+l'+1}(\pi^+\pi^-\pi^0),$$

$$C(\pi^+\pi^-\pi^0) = (\pi^-\pi^+\pi^0)$$

$$= (-1)^l(\pi^+\pi^-\pi^0).$$

So

$$CP(\pi^+\pi^-\pi^0) = (-1)^{2l+l'+1}(\pi^+\pi^-\pi^0)$$

$$= (-1)^{l'+1}(\pi^+\pi^-\pi^0).$$

Since *K* has spin 0, the most likely values of l and l' in the decay $K \to 3\pi$ are $l = l' = 0$ (in fact this is how the spin of *K* is deduced, since the Dalitz plot for $K \to 3\pi$ implies that $l = l' = 0$ — see section 8.1). In this case

$$CP(\pi^+\pi^-\pi^0)_{l=l'=0} = -(\pi^+\pi^-\pi^0)_{l=l'=0}. \tag{9-17}$$

These results are summarised in Table 3.

If *CP* conservation holds in *K* decays, as it does in beta decay, the only allowed decays are those between states with the same eigenvalue of *CP*, i.e.

Table 9.3

	P	C	CP
K_1^0	−	−	+
K_2^0	−	+	−
$(\pi^+\pi^-)_{l=0}$	+	+	+
$(\pi^+\pi^-\pi^0)_{l=l'=0}$	−	+	−

$$K_1^0 \to 2\pi, \quad K_2^0 \to (3\pi)_{l=l'=0}. \tag{9-18}$$

In particular, the decay

$$K_2^0 \to 2\pi \text{ is } \textit{strictly forbidden}. \tag{9-19}$$

Referring to Figure 1, if one looks a long way from the source, no 2π decays should be observed. Until 1964 none were, and it was believed that K decays, and in fact all weak interactions, conserve CP.

It is interesting to note that the solution to (2) is, from (18),

$$\theta^0 = K_1^0, \quad \tau^0 = K_2^0; \tag{9-20}$$

θ^0 and τ^0 are different particles, whereas θ^+ and τ^+ are the *same* particle K^+ (equation (1)). This gives a twist to the analysis of the θ-τ puzzle by Lee and Yang. The whole force of their suggestion was that θ and τ are the same particle and therefore that parity is not conserved. But we see now that although they have the same spin and parity, θ^0 and τ^0 do *not* have the same value of CP. Neither do they have the same mass, since there is a connection between mass and decay modes. Things are not so simple as we took them to be in the last chapter.

9.6 $\Delta I = \frac{1}{2}$ RULE IN $K_1^0 \to 2\pi$ DECAY

It will be recalled from Chapter 4 that the $\Delta I = \frac{1}{2}$ rule was invented to explain why the decay $K^+ \to \pi^+\pi^0$ was suppressed relative to $K_1^0 \to 2\pi$. We may employ it to calculate the relative rates of $K_1^0 \to \pi^0 + \pi^0$ and $K_1^0 \to \pi^+ + \pi^-$, and compare this with experiment

The 2π state has $l = 0$ so has a symmetric space wave function. The pions have no spin so there is no spin wave function. The generalised Pauli principle, then, demands that it has a symmetric isospin wave function, i.e. (see the remarks following (3A-28)) that $I = 0$ or 2. Since K_1^0 has $I = \frac{1}{2}$, the $\Delta I = \frac{1}{2}$ rule demands that the final (2π) state has $I = 1$ or $I = 0$. The two requirements together demand $I = 0$. From equation (3A-28) the (0, 0) state is

K MESON DECAYS AND CP VIOLATION 159

$(0, 0) = \sqrt{1/3}(\pi^+\pi^- + \pi^-\pi^+ - \pi^0\pi^0)$.

The observed decays $K_1^0 \to \pi^+\pi^-$ correspond to either combination $\pi^+\pi^-$ or $\pi^-\pi^+$, so we see that there is twice as much probability for the charged pions to belong to $I = 0$ as for the neutral pions. The decay widths Γ then obey

$$\frac{\Gamma(K_1^0 \to \pi^+\pi^-)}{\Gamma(K_1^0 \to \pi^0\pi^0)} = 2$$

Lee and Wu (see "Further Reading") point out that the agreement between various experimental determinations of this ratio is unsatisfactory, but quote a weighted average of 2.14 ± 0.10, giving moderately good support to the $\Delta I = \frac{1}{2}$ rule in $K_1^0 \to 2\pi$ decays.

9.7 THE DISCOVERY OF CP VIOLATION

In 1964 Christenson, Cronin, Fitch and Turlay (CCFT), in the course of investigating K decays to set new bounds on CP conservation, actually discovered the decay $K_2^0 \to 2\pi$, which according to (19) is forbidden if CP is conserved. The amplitude for this decay was observed to be much smaller than that for the allowed decay $K_1^0 \to 2\pi$; in fact CCFT estimated

$$\eta_{+-} \equiv \frac{\text{Amp}(K_2^0 \to \pi^+\pi^-)}{\text{Amp}(K_1^0 \to \pi^+\pi^-)} = (2.3 \pm 0.3) \times 10^{-3}$$

so the decay is comparatively rare — but it does occur, and any belief that physicists may have entertained in the fundamental nature of CP conservation is destroyed. The experiment has been repeated by the discoverers and by other workers, and there is no possible doubt that the effect is real. The decay $K_2^0 \to 2\pi^0$ has also been observed.

Before proceeding to analyse the problem, which I shall call the CP puzzle, it may be worth, by way of orientation, disposing of the most trivial "explanation", which is that what CCFT saw was not $K_2^0 \to 2\pi$, but $K_1^0 \to 2\pi$, K_1^0 being regenerated from K_2^0 in a material medium (the experiment was done in helium). CCFT in fact estimated the amount of regenerated K_1^0, and concluded that it was far too small to account for the effect. To resolve the matter finally, the experiment was repeated in a vacuum, and $K_2^0 \to 2\pi$ was still observed.

9.8 ANALYSIS OF THE CP PUZZLE

As we saw in equation (18) and from Table 3, conservation of CP allows only the decays $K_1^0 \to 2\pi$, $K_2^0 \to 3\pi$, which both occur. The CCFT observation now presents us with the *CP* puzzle

$$K_2^0 \to 2\pi \ (CP = 1)$$
$$K_2^0 \to 3\pi \ (CP = -1)$$
} "Same particle" decays into states with opposite *CP*.

It is instructive to compare this with the θ-τ puzzle

$$\theta \to 2\pi \ (P = 1)$$
$$\tau \to 3\pi \ (P = -1)$$
} "Same particle" decays into states with opposite *P*.

There were two ways out of the θ-τ puzzle (see section 8.1)

(A) θ and τ are different particles with different parity, but – "accidently" – the same spin, mass and lifetime.

(B) θ and τ are the same particle, and one of the decays violates *P*.

It is useful to analyse the *CP* puzzle in a similar (though not completely analogous) way:

1) The K_2^0s are different particles; in other words, there is a *small impurity of K_1 in the K_2 beam*, and CCFT are really seeing $K_1 \to 2\pi$. (N.B. This would *not* be due to regeneration, as explained above.)

2) The K_2^0s are the same particle and there is a *CP* violating interaction which causes the decay $K_2^0 \to 2\pi$.

Let us now consider these cases separately.

Case 1) K_1 impurity in K_2 beam

In the discussion of the K^0-$\bar{K^0}$ system, we saw that the states with a *definite lifetime* were K_1 and K_2:

$$K_1^0 = \sqrt{\tfrac{1}{2}}(K^0 - \bar{K^0}), \quad K_2^0 = \sqrt{\tfrac{1}{2}}(K^0 + \bar{K^0}).$$

Now let us *modify this slightly*, and suppose that the states with definite lifetimes are

$$K_S = \frac{1}{\sqrt{2(1 + |\epsilon|^2)}} [(1 - \epsilon)K^0 - (1 + \epsilon)\bar{K^0}] \qquad (9\text{-}21)$$

$$K_L = \frac{1}{\sqrt{2(1 + |\epsilon|^2)}} [(1 + \epsilon)K^0 + (1 - \epsilon)\bar{K^0}] \qquad (9\text{-}22)$$

K MESON DECAYS AND CP VIOLATION 161

$\epsilon \ll 1$,

where K_S is short-lived and K_L long-lived. That is,

$$K_S = \frac{1}{\sqrt{1 + |\epsilon|^2}} [K_1^0 - \epsilon K_2^0]$$

$$K_L = \frac{1}{\sqrt{1 + |\epsilon|^2}} [K_2^0 + \epsilon K_1^0];$$

the short-lived state K_S is *mostly* K_1^0, but with a *small admixture of* K_2^0. What we used to think of as $K_1 \to 2\pi$ decays were really the K_1 part of K_S decaying, and what we thought were $K_2 \to 3\pi$ decays were the K_2 part of K_L decaying: it is K_S and K_L that are the physical particles with the definite lifetimes. Now, in the CCFT observation, we have at last seen the K_1 part of K_L decaying. This is shown in Figure 3.

In a theory of this type, the blame for *CP* violation lies not on the decay $K_S \to 2\pi$ itself, but on the fact that the decaying particle is actually a mixture of particles with different values of *CP*. The question is, how can this come about? We shall return to this later.

First let us derive an expression for η_{+-}.

$$\eta_{+-} = \frac{\text{Amp}(K_2 \to \pi^+\pi^-)}{\text{Amp}(K_1 \to \pi^+\pi^-)}$$

$$= \frac{(\text{Fraction of } K_1 \text{ in } K_L) \times \text{Amp}(K_1 \to \pi^+\pi^-)}{\text{Amp}(K_1 \to \pi^+\pi^-)}$$

$$= \text{Fraction of } K_1 \text{ in } K_L$$

$$= \epsilon.$$

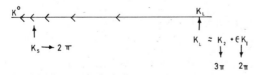

Figure 9.3 Dominant decays $K_S \to 2\pi$ and $K_L \to 3\pi$ and the rare decay $K_L \to 2\pi$

Similarly

$$\eta_{00} = \frac{\text{Amp}(K_2 \to \pi^0\pi^0)}{\text{Amp}(K_1 \to \pi^0\pi^0)}$$

$$= \epsilon,$$

i.e.

$$\eta_{+-} = \eta_{00} \qquad (9\text{-}23)$$

Case 2) CP violation in decay amplitude

There is no impurity in the K beam. The CCFT observation indicates that there is an interaction responsible for

$$K_2 \to 2\pi$$

$$CP: \; -1 \quad +1$$

which has strength

$$\frac{\text{Amp}(K_2 \to 2\pi)}{\text{Amp}(K_1 \to 2\pi)} \times G \approx 10^{-3} G$$

and violates CP. In order to derive a relation between η_{+-} and η_{00} analogous to (23) in Case (1), it is best to consider here two specific examples of theories which have been proposed[†]:

a) There is a $\Delta I = \tfrac{3}{2}$ part of the nonleptonic weak interaction, with strength $10^{-3} G$, which violates CP. In this case the decays $K_2 \to \pi^+\pi^-$ and $K_2 \to \pi^0\pi^0$ would obey a $\Delta I = \tfrac{3}{2}$ rule rather than the $\Delta I = \tfrac{1}{2}$ rule characteristic of $K_1 \to \pi^+\pi^-$ and $K_1 \to \pi^0\pi^0$. So whereas the $\Delta I = \tfrac{1}{2}$ rule gives

$$\text{Amp}(K_1^0 \to \pi^+\pi^-) = \sqrt{2}\,\text{Amp}(K_1^0 \to \pi^0\pi^0),$$

a $\Delta I = \tfrac{3}{2}$ rule would give

$$\text{Amp}(K_2^0 \to \pi^+\pi^-) = \sqrt{\tfrac{1}{2}}\,\text{Amp}(K_2^0 \to \pi^0\pi^0),$$

[†] There are actually dozens of theories of CP violation in the literature. Some are dead and many are dying. Of the remainder, I shall confine myself to citing two examples of Case (2) and one of Case (1)

K MESON DECAYS AND CP VIOLATION

so that

$$\frac{\text{Amp}(K_2^0 \to \pi^+\pi^-)}{\text{Amp}(K_1^0 \to \pi^+\pi^-)} = \frac{\tfrac{1}{2}\text{Amp}(K_2^0 \to \pi^0\pi^0)}{\text{Amp}(K_1^0 \to \pi^0\pi^0)}$$

i.e.

$$\eta_{+-} = \tfrac{1}{2}\eta_{00}. \tag{9-24}$$

b) The revolutionary suggestion has been made by Bernstein, Feinberg and Lee that, contrary to popular belief, the *electromagnetic interactions of hadrons violate C and T*, but conserve P. $K_2 \to 2\pi$ is then caused by the joint effect of weak interactions (strength G) and electromagnetic interactions (strength $\alpha/\pi \approx 2 \times 10^{-3}$), which therefore violates CP and P, according to the scheme of Table 4. They point out that it is not easy to discover if electromagnetic interactions violate C, in the sense that most reactions which would possibly show a violation are forbidden for other reasons. There is therefore not much evidence that the electromagnetic interactions of hadrons conserve C. As one test, they proposed looking for a π^+/π^- asymmetry in the decay $\eta \to \pi^+\pi^-\pi^0$. At present, the evidence is not very clear-cut (see section 2.14).

Since electromagnetic interactions obey $\Delta I = 0, 1$ (see chapter 5), combined with the $\Delta I = \tfrac{1}{2}$ weak interactions, there will be a mixture of $\Delta I = \tfrac{3}{2}$ and $\Delta I = \tfrac{1}{2}$ amplitudes. These will give, respectively, equations (23) and (24), so a mixture will give neither result; in particular

$$\eta_{00} \neq \eta_{+-}. \tag{9-25}$$

It may, of course, be that electromagnetic interactions contain a C-conserving part and a C-violating part, with the selection rules $\Delta I = 0, 1$ and $\Delta I = 0$ respectively, in which case $\eta_{00} = \eta_{+-}$. But in general we may expect (25).

Table 9.4
Symmetries of the weak (wk) and electromagnetic (em) interactions of hadrons, according to the scheme of Bernstein, Feinberg and Lee

	P	C	CP	T	CPT
wk	x	x	✓	✓	✓
em	✓	x	x	x	✓
wk + em	x	x	x	x	✓

9.9 EXPERIMENTAL RESULTS

The most exasperating property of *CP* violation, to both theorists and experimentalists, is the smallness of the effect. It is both difficult to explain and difficult to measure. Since 1964 the experimental results have varied considerably, and have only lately begun to settle down. Recent figures are[†]

$$\eta_{+-} = |\eta_{+-}|e^{i\varphi_{+-}}, \quad \eta_{00} = |\eta_{00}|e^{i\varphi_{00}}$$

$$|\eta_{+-}| = (1.95 \pm 0.03) \times 10^{-3}$$

$$|\eta_{00}| = (2.24 \pm 0.20) \times 10^{-3}$$

$$\varphi_{+-} = 43.6° \pm 3.3°$$

$$\varphi_{00} = 43° \pm 19°$$

so that, within the errors, $\eta_{+-} = \eta_{00}$ in magnitude and phase. Case (1) is favoured. Case (2a) is ruled out. The best way of testing for Case (2b) is to look for *C*-violation in other electromagnetic processes. The decay $\eta \to \pi^+\pi^-\pi^0$ has already been mentioned. In addition, the very small upper bound on the neutron electric dipole moment would seem to suggest that electromagnetic interactions conserve *C*. (See section 2.12 for a discussion of this.)

Theories belonging to Case (1) are so far consistent with experiment. It is to be noted that since in this case it is the *CP* impurity in the *K* beam which is responsible for *CP* violation, there is the important prediction that *CP violation only ever occurs in decays of neutral kaons.* Let us now turn to an example of a theory belonging to Case (1).

9.10 THE SUPERWEAK THEORY

According to this theory, due to Wolfenstein, there is an interaction which mixes K^0 and $\overline{K^0}$ and violates *CP*. In conjunction with the ordinary weak interactions, which conserve *CP* and also mix K^0 and $\overline{K^0}$, it has the effect of causing the states with a definite lifetime not to be eigenstates of *CP*. They are not K_1 and K_2 but mixtures of them. This new interaction obeys $|\Delta S| = 2$, since it mixes K^0 and $\overline{K^0}$ directly

$$K^0 \longleftrightarrow \overline{K^0}$$
$$S=1 \quad |\Delta S|=2 \quad S=-1$$

[†] See for example K. Winter, Rapporteur's talk on Weak Interactions, In *Proc. Amsterdam Conf. on Elementary Particles 1971*, (A. G. Tenner and M. J. G. Veltman, eds.) North-Holland, 1972.

K MESON DECAYS AND CP VIOLATION

Let the strength of the interaction be characterised by the coupling constant G_{swk}. The (rare) decay $K_L \to 2\pi$ is due to the *first order* mixing above, so its amplitude is proportional to G_{swk}. On the other hand the (common) decay $K_S \to 2\pi$ is due to the *second order* mixing $K^0 \leftrightarrow 2\pi \leftrightarrow \overline{K^0}$ of ordinary weak interactions, so its amplitude is proportional to G^2. So we have

$$\frac{\text{Amp}(K_L \to 2\pi)}{\text{Amp}(K_S \to 2\pi)} \approx \frac{G_{swk}}{G^2} \approx 10^{-3}$$

i.e.

$$G_{swk} \approx 10^{-3} G^2; \quad G_{swk} \approx 10^{-8} G$$

where we have put $G \approx 10^{-5} m_p^{-2}$, using the proton mass as a natural unit to compare quantities of different dimensions. The superweak interaction is then 10^8 times weaker than the ordinary weak interaction — hence the name. Apart from predicting $\eta_{+-} = \eta_{00}$, it also predicts $\varphi_{+-} = \varphi_{00} = 45°$, in agreement with experiment.

CP violation has also been discovered in the decays

$$K_L \to \pi^+ + e^- + \bar{\nu}_e, \quad K_L \to \pi^- + e^+ + \nu_e, \tag{9-26}$$

by finding a nonzero *charge asymmetry*

$$\delta = \frac{N_+ - N_-}{N_+ + N_-} \tag{9-27}$$

in the number of positrons and electrons emitted in the two decays. The experiments give $\delta = (3.22 \pm 0.29) \times 10^{-3}$, which is in agreement with the superweak theory. An interesting consequence of CP violation in these decays (26) is that they provide an *absolute distinction between matter and antimatter*; in a CP-mirror world, all the laws of physics would look the same *except* that δ would be negative. The decays (26) bear the same relation to the $K_2^0 \to 2\pi$ decays as nuclear beta decays bears to $\theta^+ \to 2\pi$, $\tau^+ \to 2\pi$; they provide a means of distinguishing matter from antimatter where the decays $K_2 \to 2\pi$ and $\theta \to 2\pi$, $\tau \to 3\pi$ merely indicated that CP (P) were violated. The situation is summarised in Table 5. The distinction between matter and antimatter provided by the decays $K_L \to \pi^+ + e^- + \bar{\nu}_e$ and $K_L \to \pi^- + e^+ + \nu_e$ is an *absolute* one because CP, as well as C and P separately, is violated.

Finally, there is a particular version of the superweak theory, proposed by Gürsey and Pais, which attributes the $|\Delta S| = 2$ K^0-$\overline{K^0}$ mixing to a spin 0 massless cosmological field with $S = 2$.

166   ELEMENTARY PARTICLES AND SYMMETRIES

Table 9.5

Symmetry	Decays which indicate symmetry is violated	Decays providing (relative or absolute) distinction between matter & antimatter (left & right)
P, C	$K^+ \to 2\pi$ $K^+ \to 3\pi$	(relative) $n \to p + e^- + \bar{\nu}_e$
CP	$K_1^0 \to 2\pi$ $K_2^0 \to 2\pi$	(absolute) $K_L \to \pi^- + e^+ + \nu_e$ $K_L \to \pi^+ + e^- + \bar{\nu}_e$

9.11 T AND CPT IN K DECAYS

Independently of CP conservation, there are tests for T invariance and CPT invariance in K decays. They take the form of equalities or non-equalities of various parameters connected with the decays. According to recent evidence[†], T invariance is definitely violated in $K_2 \to 2\pi$ decays, but CPT still holds good.

This completes our discussion of the discrete symmetry operations P, C and T and their combinations. If we accept tentatively that the superweak theory is right, we may summarise the behaviour of the various interactions under P, C and T as in Table 6. If, however, it is electromagnetism or some other mechanism which is responsible for $K_2 \to 2\pi$, the reader may modify the table accordingly. One interesting thing about the superweak theory is that if it is right there is an interaction between particles which was previously unrecognised — there are now, including gravity, 5 basic interactions.

9.12 MORE ON K MESON INTERFEROMETRY

A very interesting experiment was performed by Fitch, Roth, Russ and Vernon in 1965 which provides, by means of interferometry, a way of distinguishing matter from antimatter, over and above the distinction allowed by the positive charge asymmetry in the decays (26). The essence of the experiment is to observe the interference between the 2π decaying straight from K_L, and the 2π decaying from K_S which has been regenerated from the K_L beam:

$$\left. \begin{array}{l} K_L \to 2\pi \\ K_L \to K_S \to 2\pi \end{array} \right\} \text{observe interference}$$

[†] See, for example, K. Winter, op.cit.

K MESON DECAYS AND CP VIOLATION

Table 9.6
Basic interactions between particles, assuming there is a superweak interaction, and their invariance (✓) or non-invariance (x) under P, C, T and CPT

	P	C	T	CPT
strong	✓	✓	✓	✓
el.mag.	✓	✓	✓	✓
weak	x	x	✓	✓
superweak	x	x	x	✓

The states with definite lifetimes K_S' and K_L' in a material medium are not the same as the states K_S and K_L with definite lifetimes in a vacuum. This is because in a material $\overline{K^0}$ is (at least partially) absorbed and the mixture of K^0 and $\overline{K^0}$ is changed. The mixture with a definite lifetime will obviously depend on the material and on its density, since the denser it is, the more $\overline{K^0}$ will be absorbed. Far inside the regenerative material, K_S' will have decayed away, and pure K_L' will be left:-

$$K_L' = K_L + cK_S \tag{9-28}$$

where c is a constant depending on the nature and density of the material. The total amplitude for decay into 2π is then

$$A = \text{Amp}(K_L \to 2\pi) + c\,\text{Amp}(K_S \to 2\pi)$$

Intensity of $2\pi = |A|^2$. \hfill (9-29)

Greatest interference will be observed if the two terms in A are of equal magnitude; Fitch et al varied c by doing the experiment in different densities of material and eventually got *maximal constructive interference*.

What would happen in a world made of antimatter? K_S and K_L are the *same* in either world, since the mixture of K^0 and $\overline{K^0}$ that they contain is determined by the laws of physics, not by their surroundings. The difference in a world of antimatter is the regenerative material. It will be made of antiprotons, antineutrons and positrons, so it will be K^0 rather than $\overline{K^0}$ which is absorbed out of the original K_L beam

$$K^0 + \bar{p} \to \pi + \overline{Y}_1^*$$
$$\hookrightarrow \begin{cases} \overline{\Sigma} + \pi \\ \overline{\Lambda} + \pi \end{cases}$$

The resulting beam therefore contains more $\overline{K^0}$ rather than more K^0. Inverting (21) and (22)

$$K^0 \sim (1 - \epsilon)K_S + (1 + \epsilon)K_L$$

$$\overline{K^0} \sim -(1 + \epsilon)K_S + (1 - \epsilon)K_L.$$

We see that this is equivalent to adding K_S to the K_L beam but with opposite sign from (28):

$$K_L'' = K_L - c'K_S$$

($c' \approx c$; the difference lies in the difference between $(1 + \epsilon)$ and $(1 - \epsilon)$.) Then in place of (29) the amplitude for observing 2π is

$$A = \text{Amp}(K_L \to 2\pi) - c'\text{Amp}(K_S \to 2\pi)$$

Intensity of $2\pi = |A|^2$,

and when the interference is arranged to be greatest, by varying the density of the regenerative material, it results in *maximal destructive interference*.

Thus by use of interferometry an absolute distinction between matter and antimatter is possible from the $K \to 2\pi$ decays only.

9.13 CP VIOLATION AND THE ARROW OF TIME

Finally, since K decays are (probably) invariant under *CPT*, the fact that they violate *CP* means that they violate *T*. They define an "arrow of time"; in a time-reversed world, just as in a *CP*-mirror world, δ in (27) would be negative and the interference discussed above would be destructive. (There is, incidentally, no possibility of *communication* with a *T*-reversed world. If it were possible to send signals into and out of a region where time flows backwards, it may be possible to receive the 'reply' before sending the 'message'; in this case causality would be violated.) Thus not only is there an arrow of time associated with thermodynamics (entropy increases with time), but there is also an arrow of time on the *microscopic* scale.†

The "discrete symmetries" of nature have provided physics with a succession of surprises. As recently as 1956 it was believed that all seven operations *P, C, T, CP, CT, PT* and *CPT* were exact symmetries of nature. Within a decade it was known that at most only *CPT* is exact.

---

† There may also be an arrow of time defined by cosmology; see, for example T. Gold in *Recent Developments in General Relativity*, Pergamon Press, 1962.

FURTHER READING

K mesons

Feynman, R. P., R. B. Leighton and M. Sands, *The Feynman Lectures on Physics*, vol III, Adison-Wesley, 1965, Section 11.5

Gell-Mann, M. and A. Pais, "Behavior of Neutral Particles under Charge Conjugation", *Physical Review* **97**, 1387 (1955)

Lee, T. D. and C. S. Wu, "Weak Interactions, Chapter 9", *Annual Reviews of Nuclear Science*, **16** (1966).

Okun', L. B. *Weak Interactions of Elementary Particles*, Pergamon, 1965, Chapters 14–16.

CP violation

Bell, J. S., "Theory of Weak Interactions", in *High Energy Physics*, (C. DeWitt and M. Jacob, eds.), Gordon & Breach, 1965.

Bernstein, J., *Elementary Particles and their Currents*, Freeman, 1968, pp.302–318.

Cabibbo, N., "Possible Consequences of the $K_2^0 \to \pi^+\pi^-$ decay", in *Symmetries in Elementary Particle Physics* (A. Zichichi, ed.), Academic Press, 1965.

Christenson, J. H, J. W. Cronin, V. L. Fitch and R. Turlay, "Evidence for the 2π decay of the K_2^0 meson", *Physical Review Letters* **13**, 138 (1964).

Kabir, P. K., *The CP Puzzle: Strange Decays of the Neutral Kaon*, Academic Press, 1968.

Sachs, R. G., "Time Reversal", *Science*, **176**, 587 (1972).

Wigner, E. P. "Violations of Symmetry in Physics", *Scientific American*, December, 1965.

CHAPTER 10

The Conserved Vector Current Theory and Unitary Symmetry

10.1 INTRODUCTION – THE NEED FOR A SYMMETRY HIGHER THAN ISOSPIN

IN THE LATE 1950s and early 1960s so many new particles and resonant states were being discovered that it was becoming something of an embarrassment to particle physicists – were there really so many *elementary* particles? To anyone who believed in the ultimate simplicity of nature (and this is surely the faith of every physicist) it was uncomfortable to have to face the fact that new particles were being discovered with such inexorable regularity. Attempts were made to construct models in which all particles were "made out of" a basic few – one of these, the Sakata model, we shall have occasion to consider later – but these models, for various reasons, carried little conviction.

At the same time it was realised that the situation is ameliorated by the existence of symmetries. For example, by virtue of isospin, if we enumerate particles, we need not, for instance, count the proton and neutron separately, since they are merely states of the same particle. Similarly, there is only "one" pion, "one" Σ, etc. By utilising isospin symmetry we can reduce the number of independent particles. But around 1960 it was felt that this was not enough: there were too many isospin multiplets. What was needed was another symmetry "larger" than isospin symmetry to enable us to say of, for instance the multiplets N, Σ, Λ and Ξ, "these are all states of the same particle". As it stands, of course, this statement is implausible; even if we ignore electromagnetic and weak interactions, there is still a lot of difference between N, Σ, Λ and Ξ. But consider the statement "N, Σ, Λ and Ξ are *approximately* states of the same particle". This at least is not *so* implausible. They have the same spin and parity, and their masses are not very widely spaced – see for instance Figure 8-2. This statement is in fact one of the conclusions of the "higher" symmetry proposed by Gell-Mann. It is called *unitary symmetry*. It is the purpose of this chapter and the next one to describe this very important theory of Gell-Mann's.

There were, around 1960, many attempts to discover a higher symmetry. Most workers approached the problem within the context of strong interactions only. Gell-Mann however, approached it also through the "back door" of weak interactions, generalising an idea which he and Feynman had proposed in 1958. In this

THE CONSERVED VECTOR CURRENT THEORY

chapter we shall concentrate on the approach through the weak interactions, and in the next one on the approach through strong interactions. It is one of the most remarkable facts that they lead to the same symmetry scheme. The reason for this is not completely understood, although Gell-Mann has pioneered some interesting investigations.† ‡

10.2 NEUTRON DECAY AND MUON DECAY

Let us very briefly rehearse one or two facts about nuclear beta decay.§ According to Fermi, it is due to an interaction which causes the transition $n \to p + e^- + \bar{\nu}$. *Allowed* transitions are of two types; Fermi and Gamow-Teller. In *Fermi transitions* the spin of the nucleus is unchanged. This means that $(e^- \bar{\nu})$ have $J = 0$, and since $L = 0$ for allowed transitions, then $S = 0$ and we have (where the arrows denote spin up and spin down states)

Fermi transition: $n \to p + e^- + \bar{\nu}$
$\quad\quad\quad\quad\quad\quad\;\; \uparrow \quad \uparrow \quad\; \uparrow \quad \downarrow$
$\quad\quad\quad\quad\quad\quad\;\; \underbrace{\quad\quad} \;\; \underbrace{\quad\quad}$
$\quad\quad\quad\quad\quad\quad\;\; \Delta J=0 \quad\;\; S=0$ \hfill (10-1)

The transition rate for these processes is

$$T = \frac{2\pi}{\hbar} G_F^2 |\int (\psi_p^* \psi_n)(\psi_e^* \psi_\nu) d\tau|^2 \rho(E)$$ (10-2)

where G_F is called the Fermi weak coupling constant and $\rho(E)$ is the density of final states.

In *Gamow-Teller transitions* the nuclear spin changes;

Gamow–Teller transition: $n \to p + e^- + \bar{\nu}$
$\quad\quad\quad\quad\quad\quad\quad\quad\;\; \uparrow \quad \downarrow \quad\; \uparrow \quad \uparrow$
$\quad\quad\quad\quad\quad\quad\quad\quad\;\; \underbrace{\quad\quad} \;\; \underbrace{\quad\quad}$
$\quad\quad\quad\quad\quad\quad\quad\quad\;\; \Delta J=1 \quad\;\; S=1$ \hfill (10-3)

---

† See for example M. Gell-Mann in *Proceedings of the Eleventh International Universitätswochen für Kernphysik,* Schladming, Austria (ed. P. Urban), Springer 1972, page 733.

‡ Unitary symmetry was proposed independently by M. Gell-Mann and Y. Ne'eman. Gell-Mann, however, spelled out the consequences in greater detail. For this and for his other contributions to the theory of elementary particles, Gell-Mann was awarded the Nobel prize in physics in 1969.

§ For a more complete account, see H. A. Enge, *Introduction to Nuclear Physics,* Addison-Wesley, 1966, pp.313 et seq.: or R. B. Leighton, *Principles of Modern Physics,* McGraw-Hill, 1959, section 15.8.

and the transition rate is

$$T = \frac{2\pi}{\hbar} G_{GT}^2 |\int (\psi_p^* \sigma \psi_n)(\psi_e^* \sigma \psi_\nu) d\tau|^2 \rho(E) \tag{10-4}$$

where G_{GT} is the Gamow–Teller coupling constant and σ is the spin operator.

By studying pure Fermi and pure Gamow–Teller decays the constants G_F and G_{GT} can be found from experiment to be

$$G_F = (1.415 \pm 0.002) \times 10^{-49} \text{ erg cm}^3$$

$$G_{GT} = (1.67 \pm 0.04) \times 10^{-49} \text{ erg cm}^3$$

$$G_{GT} = (1.18 \pm 0.03) G_F. \tag{10-5}$$

The expressions $(\psi^*\psi)$ and $(\psi^*\sigma\psi)$ occurring in (2) and (4) above are not relativistically covariant. $\psi^*\psi$ is not a scalar, but the fourth component of a 4-vector, and $\psi^*\sigma\psi$ are 3 components of a 4-pseudovector (since spin σ is a pseudovector). In the Dirac theory the relativistic expressions are

$$\text{Vector } V_\lambda = \bar{\psi}\gamma_\lambda \psi = (\bar{\psi}\gamma\psi, i\bar{\psi}\gamma_4\psi) \xrightarrow[\text{limit}]{\text{non-rel.}} (0, i\psi^*\psi) \tag{10-6}$$

$$\text{Axial Vector Current } A_\lambda = \bar{\psi}\gamma_\lambda\gamma_5 \psi = (\bar{\psi}\gamma\gamma_5\psi, i\bar{\psi}\gamma_4\gamma_5\psi) \xrightarrow[\text{limit}]{\text{non-rel.}} (\psi^*\sigma\psi, 0) \tag{10-7}$$

whose nonrelativistic limits are the Fermi and Gamow–Teller expression. Here ψ is a 4-column vector, $\bar{\psi}$ is related to ψ^*, and the γs are 4×4 matrices.[†] The suffix λ takes on the values $1, \ldots, 4$. In a completely relativistic theory, then, instead of the expressions (2) and (4), there should be the "current-current" interactions

Fermi:
$$(\psi_p^*\psi_n)(\psi_e^*\psi_\nu) \to \sum_{\lambda=1}^{4} (\bar{\psi}_p\gamma_\lambda\psi_n)(\bar{\psi}_e\gamma_\lambda\psi_\nu) \tag{10-8}$$

Gamow–Teller:
$$(\psi_p^*\sigma\psi_n) \cdot (\psi_e^*\sigma\psi_\nu) \to \sum_{\lambda=1}^{4} (\bar{\psi}_p\gamma_\lambda\gamma_5\psi_n)(\bar{\psi}_e\gamma_\lambda\gamma_5\psi_\nu). \tag{10-9}$$

[†] For an introduction to the Dirac theory, see Leighton, op.cit., section 20-11.

THE CONSERVED VECTOR CURRENT THEORY

Some decays are pure Fermi, and some pure Gamow–Teller, but free neutron decay, as may be seen from (1) and (3), may be either. In general, then, we may expect *both* terms to be present. Note, however, that V_λ and A_λ, the vector and axial vector currents, have *opposite parity*, so if both are present, parity will be violated in neutron decay. This, as we know, is the case.

The simplest hypothesis for the interaction responsible for neutron decay is that it involves the vector and axial vector currents in equal portions

$$H = \sum_{\lambda=1}^{4} G_F(V_\lambda^{pn} + A_\lambda^{pn})(V_\lambda^{ev} + A_\lambda^{ev}) = \sum_{\lambda=1}^{4} G_F\left\{(\bar\psi_p\gamma_\lambda(1+\gamma_5)\psi_n)(\bar\psi_e\gamma_\lambda(1+\gamma_5)\psi_\nu)\right\}$$
(10-10)

and the transition rate is then

$$T = \frac{2\pi}{\hbar} \tfrac{1}{2} |\int Hd\tau|^2 \rho(E)$$
(10-11)

(10) is known as the *V-A* theory. The factor of ½ appears in (11) to compensate for the doubling of terms introduced by $(1 + \gamma_5)$. Since the vector currents give the Fermi transitions and the axial vector currents the Gamow–Teller transitions, their symmetrical appearance in (10) implies that

$$G_F = G_{GT}$$

which is wrong – see equation (5). On the other hand, the two component neutrino theory demands that the (*ev*) term contain *V* and *A* symmetrically. (This is because the matrix $(1 + \gamma_5)$ selects the left-handed neutrino and right-handed antineutrino only.) We may then amend (10) to give the more realistic interaction

$$H = \sum_{\lambda=1}^{4} G_F (\bar\psi_p\gamma_\lambda(1+1.18\gamma_5)\psi_n)(\bar\psi_e\gamma_\lambda(1+\gamma_5)\psi_\nu)$$
(10-12)

which has taken into account the fact that $G_{GT} = 1.18\, G_F$, as in (5). We may call (12) the V − 1.18A theory

Now consider μ decay

$$\mu^- \rightarrow e^- + \bar\nu_e + \nu_\mu$$
(10-13)

whose lifetime is

$$\tau = (2.198 \pm 0.001) \times 10^{-6} \text{ sec.}$$
(10-14)

Analogously to (10) and (11) for neutron decay the interaction energy is

$$H_\mu = \sum_{\lambda=1}^{4} G_F (\bar\psi_\mu \gamma_\lambda (1+\gamma_5)\psi_\nu)(\bar\psi_e \gamma_\lambda (1+\gamma_5)\psi_\nu),\tag{10-15}$$

leading to a transition rate

$$T = \frac{2\pi}{\hbar} \tfrac{1}{2} |\int H_\mu d\tau|^2 \rho(E) \tag{10-16}$$

where the symmetric appearance of V and A in both terms in (15) is due to the fact that both terms contain neutrinos, which are believed to be massless and two-component. Using (16), the lifetime of μ is calculated to be

$$\frac{1}{\tau} = \frac{G_\mu^2 m_\mu^5}{192\pi^3}$$

and comparison with the experimental value (14) gives

$$G_\mu = (1.4350 \pm 0.003) \times 10^{-49} \text{ erg cm}^3. \tag{10-17}$$

It is striking to notice how close are the values of G_μ and G_F, comparing (5) and (17). They are almost the same, but unfortunately, they are not the same within the errors. If *all* the coupling constants G_F, G_{GT} and G_μ were equal, there would be a *universality* in weak interactions; that is, all decays would be of the same strength. However, while universality is hinted at by (5) and (17), it is nevertheless firmly denied. From (5) and (17)

$$G_\mu \simeq G_F: \quad \frac{G_\mu - G_F}{G_F} = 2.8 \pm 0.13\%, \tag{10-18}$$

$$G_{GT} = 1.18 G_F. \tag{10-19}$$

Two questions then arise;

(1) Why are G_μ and G_F so close?

(2) Why is $G_{GT}/G_F = 1.18$? \hfill (10-20)

The second question will be taken up in Chapter 12. Let us now turn our attention to question (1).

10.3 THE CONSERVED VECTOR CURRENT THEORY

For the present it is simplest to ignore the discrepancy between G_μ and G_F, and suppose that they are essentially equal. That is, the (Fermi) coupling of the (pn) pair to the $(e\nu)$ field is of the same strength as the coupling of $(\mu\nu)$ to $(e\nu)$. Simple as it sounds, however, it turns out that this statement has very non-trivial implications. To help us to see this, consider the very similar observation that p and μ^+ have the same electric charge

$$Q(p) = Q(\mu^+) = e. \qquad (10\text{-}21)$$

The charge on a particle determines its coupling to the electromagnetic field (photons). For a muon, this coupling is simple enough, as shown in Figure 1. A proton, however, while it is sometimes just a proton (Figure 2(a)), has an amplitude for virtual dissociation into $n + \pi^+$ (Figure 2(b)) and indeed the π^+ may further dissociate into, say, $\Sigma^+ + \bar{\Lambda}$ (Figure 2(c)). These dissociations due to the virtual pion cloud have the effect of *screening* the proton from the photon. This would be expected to make a difference to the charge, but according to (21) it makes no difference. The reason is that *charge is conserved*; no matter what dissociations the proton may undergo, because the charge is never lost, the photon always sees the same charge, irrespective of what particle may be carrying it, so the screening does not matter. The equation for charge conservation is

$$\frac{dQ}{dt} = 0 \qquad (10\text{-}22)$$

and this follows from the equation for *electromagnetic current conservation*

$$\sum_\lambda \frac{\partial}{\partial x_\lambda} J_\lambda = \frac{\partial}{\partial x} \cdot \mathbf{J} + \frac{\partial}{\partial t} J_0 \qquad (10\text{-}23)$$

where $J_4 = iJ_0$ and Q is related to J_0. (In fact by integrating (23) over all space, and assuming that $\mathbf{J} \to 0$ at infinity, (22) follows if $Q = \int J_0 d^3x$.)

Figure 10.1 Feynman diagram for the coupling of a photon to a muon

176   ELEMENTARY PARTICLES AND SYMMETRIES

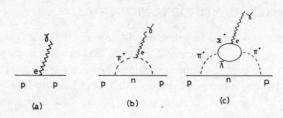

Figure 10.2 Feynman diagrams for some of the ways in which a photon couples to a proton

Similar considerations hold for the (near-)equality (18). G_μ is measured in μ decay, according to the Feynman diagram of Figure 3. In n decay, however, the existence of the pion cloud means that besides the "bare" decay of Figure 4(a), it may be a virtual pion, or a virtual Σ, or indeed anything else, which decays by interacting with the $(e\nu)$ field. The equality (18) can only hold to all orders if this screening of the bare (np) pair does not matter; in other words, if the "weak charge" G_F is *conserved*:

$$\frac{dG_F}{dt} = 0. \tag{10-24}$$

This means that *all* the (Fermi) transitions

$$\pi^- \to \pi^0 + e^- + \bar{\nu}_e$$

$$\Sigma^- \to \Sigma^0 + e^- + \bar{\nu}_e \tag{10-25}$$

etc.

are characterised by the *same* coupling constant G_F. In addition, conservation of the weak charge G_F is equivalent to conservation of the "Fermi" current, which, as we saw above, is the vector current V_λ, so we finish up with a *conserved vector current* (CVC)

$$\frac{\partial}{\partial x_\lambda} V_\lambda = \frac{\partial}{\partial \mathbf{x}} \cdot \mathbf{V} + \frac{\partial}{\partial t} V_0 = 0. \tag{10-26}$$

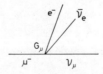

Figure 10.3 Feynman diagram for muon decay

THE CONSERVED VECTOR CURRENT THEORY 177

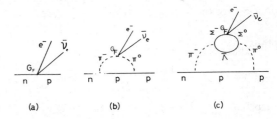

Figure 10.4 Feynman diagrams for some of the ways in which the neutron decays

(Again, (24) may be derived from (26) by integrating over all space, dropping the surface terms, and putting $G_F = \int V_0 d^3x$.)

The fact that the Fermi transitions (25) are of the same strength as the Fermi transition in nuclear beta decay provides a test of the CVC theory. In fact, the decay

$$\pi^- \to \pi^0 + e^- + \bar{\nu}_e \tag{10-27}$$

is a pure Fermi transition ($0^- \to 0^-$), so its strength is given uniquely by G_F. The lifetime is then predicted to be

$$\tau = 2.49 \pm 0.03 \text{ sec.}$$

This should be compared with the lifetime of ordinary π^- decay

$$\tau_{\pi^-} = (2.55 \pm 0.03) \times 10^{-8} \text{ sec.}$$

The branching ratio R is

$$R = \frac{\tau_{\pi^-}}{\tau} = (1.02 \pm 0.02) \times 10^{-8}.$$

In other words, about one pion in one hundred million decays according to (27). The experimental for R is

$$R_{\exp} = (1.3 \pm 0.2) \times 10^{-8},$$

which is consistent with the theoretical figure, considering the errors involved.

Now the question arises, *why* is the vector current conserved? A very suggestive answer was advanced by Gershtein and Zeldovich, and later by Feynman and Gell-Mann. To appreciate its significance, consider the following facts about the vector current V_λ.

1) It is responsible for the hadronic decays

$n \to p + \text{leptons}$

$\pi^- \to \pi^0 + \text{leptons}$

$\Sigma^- \to \Sigma^0 + \text{leptons}$ \hfill (10-28)
etc

which have the selection rules

$\Delta I_3 = 1, \quad \Delta S = 0$ \hfill (10-29)

2) Consider the vector current $V_\lambda^{\pi^- \pi^0}$ for the pion decay $\pi^- \to \pi^0 + e^- + \bar{\nu}_e$. $V_\lambda^{\pi^- \pi^0}$ must be a function of the pion wave functions, and must also be a 4-vector. The only 4-vector associated with the pions is their energy-momentum, so we put

$$V_\lambda^{\pi^- \pi^0} = \varphi_{\pi^-}^* [p_\lambda(\pi^-) + p_\lambda(\pi^0)] \varphi_{\pi^0}. \tag{10-30}$$

Assuming the pion wave functions are plane waves, their dependence on $x_\lambda = (\mathbf{x}, it)$ is

$$\varphi_\pi \propto \exp i(\mathbf{p} \cdot \mathbf{x} - Et) = \exp i \sum_{\lambda=1}^{4} p_\lambda x_\lambda. \tag{10-31}$$

In this case

$$\frac{\partial}{\partial x_\lambda} \varphi_{\pi^0} = i p_\lambda(\pi^0) \varphi_{\pi^0}$$

$$\frac{\partial}{\partial x_\lambda} \varphi_{\pi^-}^* = -i p_\lambda(\pi^-) \varphi_{\pi^-}^*.$$

The divergence of $V_\lambda^{\pi^- \pi^0}$ is then

$$\sum_\lambda \frac{\partial}{\partial x_\lambda} V_\lambda^{\pi^- \pi^0} = \frac{1}{2} \sum_\lambda \left\{ \left(\frac{\partial}{\partial x_\lambda} \varphi_{\pi^-}^* \right) (p_\lambda(\pi^-) + p_\lambda(\pi^0)) \varphi_{\pi^0} + \varphi_{\pi^-}^* (p_\lambda(\pi^-) + p_\lambda(\pi^0)) \left(\frac{\partial}{\partial x_\lambda} \varphi_{\pi^0} \right) \right\}$$

THE CONSERVED VECTOR CURRENT THEORY   179

$$= i \sum_\lambda \{-p_\lambda(\pi^-)p_\lambda(\pi^-) + p_\lambda(\pi^0)p_\lambda(\pi^0)\} \times \varphi^*_{\pi^-}\varphi_{\pi^0}$$

$$= i\{m^2(\pi^-) - m^2(\pi^0)\}\varphi^*_{\pi^-}\varphi_{\pi^0}. \qquad (10\text{-}32)$$

The condition that the current is *conserved*, (26), is then

$$m(\pi^-) = m(\pi^0). \qquad (10\text{-}33)$$

The selection rules (29) are those obeyed by the *isospin raising operator* I_+ — see for example Figures 3-4 and 3-10 (49 and 57). Also condition (33) is one that holds in the limit in which *isospin is conserved*. This suggests that we identify

vector current for weak interactions ↔ isospin current

weak charge $G_F \leftrightarrow I_+, I_-$. (10-34)

The isospin current is (approximately) conserved since isospin is an (approximate) symmetry. Identifying the weak vector current with the isospin current then provides an explanation of why the vector current is conserved. The conserved weak charge G_F in processes like (26) is then identified with the isospin raising operator I_+, and the corresponding weak charge in the (endothermic) reactions $p \to n$ + leptons, $\pi^0 \to \pi^-$ + leptons, with I_-.

10.4 SUMMARY OF CVC AND STEPS TOWARDS UNITARY SYMMETRY

For clarity, it is best to summarise the steps outlined above.

a) It is observed that the strengths of μ decay and the Fermi transitions in nuclear beta decay are almost the same, $G_\mu \simeq G_F$.

b) Even if a relation $G_\mu \simeq G_F$ holds for the "bare" coupling constants, the actual *physical* value of G_F will in general be considerably different, because of the screening effect of the pion cloud.

c) The condition for the effect of screening to be irrelevant (as they are for the proton charge), is that the weak charge G_F — and correspondingly the weak vector current V_λ — is *conserved*. This is the CVC theory.

d) The current V_λ is *automatically* conserved if it is identified with the isospin current. This is because isospin, being a symmetry of the strong interactions, is (almost) conserved; as evidence to support this identification, note that the selection rules of the $\Delta S = 0$ weak interactions, and therefore of the weak charge, are $\Delta I_3 = \pm 1$ — these are exactly the selection rules of the I-spin raising and lowering operators I_+ and I_-.

180 ELEMENTARY PARTICLES AND SYMMETRIES

The reader has by now probably forgotten that the main aim of this chapter, as outlined in the introduction, is to deduce a symmetry higher than isospin, so as to incorporate strangeness. For this purpose the significant part of the summary above is (d). In very general terms it states that there is a connection between the selection rules of the $\Delta S = 0$ weak leptonic decays, and the strong interaction symmetry — isospin — which does not involve strangeness. This connection is that the generators of the symmetry (I_+, I_-) obey the selection rules of the weak transitions.

Gell-Mann used this as a guideline to deduce a higher symmetry. He supposed that the raising and lowering operators of the symmetry were given by the selection rules of all leptonic weak decays — strangeness changing as well as strangeness conserving. These selection rules are

$$\Delta S = 0, \quad \Delta I_3 = \pm 1$$

$$\Delta S = \pm 1, \quad \Delta I_3 = \pm \tfrac{1}{2}$$

which are equivalent to (5-19) and (5-35). They are shown in Figure 5. The raising and lowering operators corresponding to $\Delta S = 0$, $\Delta I_3 = \pm 1$ are simply I_+ and I_-. Calling the raising and lowering operators corresponding to $\Delta S = \pm 1$, $\Delta I_3 = \pm \tfrac{1}{2}$ V_+ and V_-, we have

$$(\Delta S = 0, \Delta I_3 = \pm 1) \leftrightarrow I_+, I_-$$

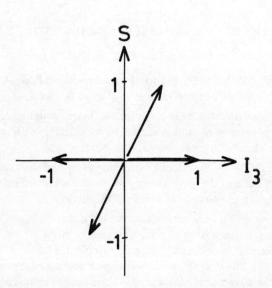

Figure 10.5 The leptonic weak interaction selection rules

THE CONSERVED VECTOR CURRENT THEORY

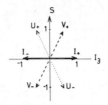

Figure 10.6 The raising and lowering operators for unitary symmetry

$(\Delta S = \pm 1, \Delta I_3 = \pm \tfrac{1}{2}) \leftrightarrow V_+, V_-$

as shown in Figure 6. I_+, I_- and I_3 have the commutation relations

$$[I_1, I_2] = iI_3 \text{ \& cyclic} \tag{10-35}$$

where $I_\pm = I_1 \pm iI_2$. These imply, for example, that

$$[I_+, I_-] = 2I_3. \tag{10-36}$$

It was pointed out in Chapter 2 that symmetry transformations, and therefore the generators of the transformations, form a *group*. In Chapter 3 (see especially the remarks following (3-16)) we saw that the group of isospin transformations is simply the three dimensional rotation group, denoted SO_3 or SU_2. Its generators are I_1, I_2 and I_3, or equivalently, I_+, I_- and I_3. This group is now *enlarged* by the addition of V_+ and V_- above. In fact, V_+, V_- and V_3 are the generators of another rotation group SU_2 called *V-spin*

$$[V_1, V_2] = iV_3 \text{ \& cyclic} \tag{10-37}$$

where

$V_\pm = V_1 \pm iV_2.$

However, V-spin and I-spin do not commute, for example

$V_+ I_- \neq I_- V_+$

$[V_+, I_-] \neq 0.$

This means that I-spin and V-spin, which separately generate rotation groups, when taken together do not generate a group. Other generators must be added,

182 ELEMENTARY PARTICLES AND SYMMETRIES

and these are called U_+, U_- and U_3, the generators of *U-spin* (see Figure 6). We then have

$$[U_1, U_2] = iU_3 \ \& \ \text{cyclic} \tag{10-38}$$

where

$$U_\pm = U_1 \pm iU_2.$$

These 9 generators of *I*-spin, *V*-spin and *U*-spin, taken together form a closed system and generate a group SU_3. (Actually SU_3 has only 8 generators, since one relation holds between the 9.)

The operators U_+ and U_- do not correspond to the selection rules for any observed weak leptonic decays — they merely have to be added for completeness. Using these operators for *I*-spin, *V*-spin and *U*-spin (so named to resemble the conjugation of verbs!), we now see what symmetry of the strong interactions they generate.

10.5 UNITARY SPIN SUPERMULTIPLETS: THE EIGHTFOLD WAY

Let us start with the proton p, plotted on a diagram, Figure 7, of Y against I_3 (where $Y = S + B$.). Let us suppose that p is the "highest state" in the supermultiplet, just as it is in its own isospin multiplet, Then as a generalisation of $I_+ p = 0$ (equation 3-11)), we have

$$I_+ p = V_+ p = U_+ p = 0. \tag{10-39}$$

To get other particles, we must apply the remaining lowering operators I_-, V_- and U_-. I_- gives the neutron as before, and V_- and U_- give particles with $I_3 = 0$,

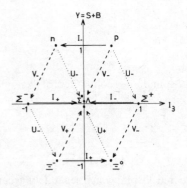

Figure 10.7 The eight $\tfrac{1}{2}^+$ baryons in the unitary spin supermultiplet (baryon octet)

THE CONSERVED VECTOR CURRENT THEORY   183

$Y = 0$ and $I_3 = 1$, $Y = 0$ respectively. These may be identified with Λ or Σ^0, and Σ^+, respectively. We may now apply V_- and U_- to n, and obtain particles with $I_3 = -1$, $Y = 0$ and $I_3 = 0$, $Y = 0$, which we identify with Σ^- and Σ^0 or Λ, respectively. We carry on in this way until we get a "closed" pattern; that is, a symmetrical pattern with a "lowest state" of opposite quantum numbers to the "highest state". In this case, since p ($I_3 = \frac{1}{2}$, $Y = 1$) is the highest state, obeying (39), then Ξ^- ($I_3 = -\frac{1}{2}$, $Y = -1$) is the lowest state, and obeys

$$I_- \Xi^- = U_- \Xi^- = 0.$$

We now have a "supermultiplet" of the 8 particles $p, n, \Sigma^+, \Sigma^0, \Sigma^-, \Lambda, \Xi^-$, with Σ^0 and Λ appearing at the same point $I_3 = Y = 0$.† *All these particles have $B = 1$ and $J^P = \frac{1}{2}^+$*; that is, they differ only in their isospin and strangeness. This supermultiplet, or superfamily, is a family of 4 isospin families with quantum numbers and masses as follows

N $I = \frac{1}{2}$ $Y = 1$ ($S = 0$)   939 MeV

Λ $I = 0$ $Y = 0$ ($S = -1$)   1115 MeV

Σ $I = 1$ $Y = 0$ ($S = -1$)   1192 MeV

Ξ $I = \frac{1}{2}$ $Y = -1$ ($S = -2$)   1314 MeV.

The masses are not the same, but as a measure of their discrepancy,

$$\frac{m(\Sigma) - m(N)}{m(\Sigma) + m(N)} = \frac{253}{2131} \approx 12\%$$

$$\frac{m(\Xi) - m(N)}{m(\Xi) + m(N)} = \frac{375}{2253} \approx 17\%, \tag{10-40}$$

so we may reasonably say that the particles are *approximately* degenerate.

Experimentally these are the only particles with $J^P = \frac{1}{2}^+$ and $B = 1$ in this mass range, and this higher symmetry has gathered up all these particles into one supermultiplet. We may say that we have a sort of compound quantum number called *unitary spin* which is a generalisation of isospin to include isospin and strangeness. The eight particles N, Σ, Λ and Ξ are eight states of *one* particle with $J^P = \frac{1}{2}^+$,

---

† The reader might object that we need only have 7 particles, with Λ not appearing. To prove that there must be 8 requires a proper treatment by group theoretic methods; it will, however, be made more plausible in the next chapter.

$B = 1$ and mass around 1100 MeV. We can imagine a hypothetical limit in which all 8 particles would be degenerate. We now say, with Gell-Mann, that

strong interactions are approximately invariant under rotation in unitary spin space. Hadrons therefore exist in superfamilies (supermultiplets) whose members form an approximately degenerate family of isospin multiplets, with differing I and S, but are otherwise identical

The rotations in unitary spin space are rotations which in general transform any one particle in the supermultiplet of Figure 7 into a *mixture* of other particles. Strong interactions are invariant under these rotations if they do not distinguish the ½⁺ baryons one from another. The philosophy of unitary symmetry is that this is approximately the case.

Since there are eight baryons in the supermultiplet, we talk about a baryon *octet*. Gell-Mann emphasised that in this eightfold way, since the eight particles are merely eight states of one particle, then no one of them is more elementary that any other one — they are to be treated on an equal footing. This was in contrast to the model of Sakata, who viewed p, n and Λ as being truly elementary, and the other baryons as made out of combinations of these three and their antiparticles: for example, $\Sigma^+ = (p\bar{n}\Lambda)$. The eightfold way and the Sakata model are incompatible.

When we turn to the 0^- mesons, we see that there also 8 of them. Moreover they fit into *exactly* the same pattern of quantum numbers as do the ½⁺ baryons (as long as we plot Y (and not S) against I_3):

K	$I = ½$	$Y = S = 1$	495 MeV
η	$I = 0$	$Y = S = 0$	549 MeV
π	$I = 1$	$Y = S = 0$	140 MeV
$\bar{K}$	$I = ½$	$Y = S = -1$	495 MeV

and are plotted in Figure 8. Apart from I and S they all have identical quantum numbers, $J^P = 0^-$, $B = 0$. Their masses however are more problematic:

$$\frac{m(K) - m(\pi)}{m(K) + m(\pi)} = \frac{355}{635} \approx 56\%. \tag{10-41}$$

This makes it less convincing to say that the isospin families are "approximately degenerate". The situation, however, is not so bad as it may seem, and more will be said about the pseudoscalar meson masses in the next chapter and in Chapter 12.

THE CONSERVED VECTOR CURRENT THEORY

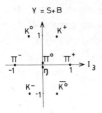

Figure 10.8   The 0^- meson octet

10.6 UNITARY SPIN SUPERMULTIPLETS: THE TENFOLD AND ONEFOLD WAYS

Is the pattern of Figure 7 the only possible pattern that can be formed from $I_\pm$, $U_\pm$ and $V_\pm$? The answer is obviously no; just as $I_\pm$ can generate families with isospin $I = 0, \frac{1}{2}, 1, \ldots$, so in general we may expect many different arrangements of supermultiplets to be generated by $I_\pm$, $V_\pm$ and $Y_\pm$ in Figure 6. On the other hand, although *in principle* any of the isospins $0, \frac{1}{2}, 1, \frac{3}{2}, 2, \frac{5}{2}, \ldots$, is possible, *in fact* the only ones observed are $I = 0, \frac{1}{2}, 1, \frac{3}{2}$. It turns out that a similar situation exists for unitary spin; only three of all possible supermultiplet patterns are at present believed to exist in nature. One of these is the octet, illustrated above. The other two contain 10 particles and 1 particle, and are called the *decuplet* and *singlet* supermultiplets.

Let us first consider the decuplet. This corresponds to the pattern in Figure 9. It is made up of the following isospin multiplets

$I = \frac{3}{2}$ $Y = 1$ (4 particles)

$I = 1$ $Y = 0$ (3 particles)

$I = \frac{1}{2}$ $Y = -1$ (3 particles)

$I = 1$ $Y = -2$ (1 particle).

A set of known particles which fits into a decuplet is the set of $J^P = \frac{3}{2}^+$ baryon resonances. In 1962 the relevant known particles were

$\Delta(1236 \text{ MeV})$ $I = \frac{3}{2}$ $Y = 1$ $(S = 0)$

$Y_1^* = \Sigma(1385 \text{ MeV})$ $I = 1$ $Y = 0$ $(S = -1)$

$\Xi^* = \Xi(1530 \text{ MeV})$ $I = \frac{1}{2}$ $Y = -1$ $(S = -2)$

and these 9 particles, it was observed, occupy the "top" 9 places of the decuplet. This means the tenth particle must exist with $I = 0$, $Y = -2$ ($S = -3$) and $J^P = \frac{3}{2}^+$. Moreover, although the above particles do not have the same mass, it may be noticed from above that the masses are *equally spaced* with respect to Y, with a gap of 146 MeV. If we assume that this equal spacing holds good for the missing particle, called Ω^-

$$m(\Omega) - m(\Xi^*) = m(\Xi^*) - m(Y_1^*) = m(Y_1^*) - m(\Delta) \tag{10-42}$$

then $m(\Omega) = 1676$ MeV. Since it has $S = -3$, the lowest mass decay mode which conserves strangeness is

$$\Omega \rightarrow \Xi^0 + K^-$$
mass (MeV) 1676 1315   498

and this is forbidden energetically. So Ω^- is *stable* with respect to the strong interactions, and is probably the last remaining stable particle (whence the name, Ω being the last letter of the Greek alphabet). Ω^- is then predicted to decay via $\Delta S = 1$ weak interactions into

$$\Omega^- \rightarrow \begin{cases} \Lambda + K^- \\ \Xi^- + \pi^0 \\ \Xi^0 + \pi^- \end{cases}$$

and since the decay is weak, the lifetime is about 10^{-10} sec, and the track should be visible in a bubble chamber. This prediction of the *exact* properties of a particle is, to say the least, very unusual in elementary particle physics, and the discovery of Ω^- in 1964 with a mass of 1686 ± 12 MeV was a great triumph for unitary symmetry.

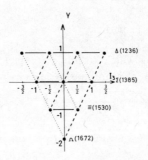

Figure 10.9 The decuplet of $\frac{3}{2}^+$ particles, with the raising and lowering operators of Figu

THE CONSERVED VECTOR CURRENT THEORY 187

Finally, there is the supermultiplet of one particle, the singlet. It is simply a particle with $I = Y = 0$, i.e. a baryon with $I = 0, S = -1$ or a meson with $I = S = 0$. There are some particles believed to be singlets, and they will be discussed in the next chapter.

10.7 REMARKS ON THE MEANING OF UNITARY SYMMETRY

It will be recalled that the motive for searching for a symmetry higher than isospin was to alleviate the problem of the large number of particles which exist. If there is a symmetry, this means there are fewer *independent* particles. Now we have seen that a higher symmetry does exist; particles are not really independent, for they not only belong to isospin multiplets, but the isospin multiplets belong to unitary spin supermultiplets, which contain several isospin multiplets with differing strangeness (hypercharge). This unitary symmetry, or SU_3 symmetry, is more difficult to appreciate than isospin symmetry, so let us reconsider some of its features.

a) *All hadrons belong to SU_3 supermultiplets*
This is axiomatic to the theory. Hadrons by definition experience strong interactions, and strong interactions conserve unitary spin, so hadrons must belong to SU_3 supermultiplets; in the language of group theory, hadrons must belong to "representations" of SU_3. This is a very strong statement; if we discover an isospin multiplet, we know it must belong to an SU_3 supermultiplet, so we *automatically predict more particles*. For instance, if we had discovered p and n, and suspected they belonged to an octet, we should automatically predict Σ^+, Σ^0, Σ^-, Λ, Ξ^0 and Ξ^-, all with $J^P = \frac{1}{2}^+$ and similar mass. Of course, p and n may not belong to an octet, but in SU_3 there is *no* supermultiplet containing only two particles –apart from 1, the minimum number of particles is 3. (The question of what supermultiplets may exist will be considered in the next chapter.) So, having discovered p and n, we should in any case predict *something* else; exactly what depends on what supermultiplet p and n belong to. This is the sense in which the particles are not independent – only the supermultiplets are.

b) *SU_3 is an approximate ("broken") symmetry*
The isospin multiplets within a supermultiplet are not degenerate. We saw that for the $\frac{1}{2}^+$ baryon octet, $\Delta m/m \approx 15\%$, and for the 0^- meson octet, $\Delta m/m \approx 56\%$. For the baryon decuplet, $\Delta m/m \approx 9\%$. Ignoring for the moment the problematic 0^- meson octet, we may say the symmetry holds to within about 15%. Because of its approximate nature, it is sometimes called a *"broken" symmetry*.

The question arises, why is the symmetry broken; and what causes the breaking? We know for instance that in the case of isospin it is probably electromagnetism that causes the breaking. In the case of unitary spin symmetry, however, there is no known force in nature which could be responsible for the breaking. Ne'eman has conjectured that such an interaction may exist, and may also be responsible for the mysterious μ-e mass difference.

In any case it is convenient to represent the broken symmetry by splitting up the strong interaction Hamiltonian:

$$H_{\text{strong}} = H_{SU_3} + \epsilon H_{\text{medium strong}} \tag{10-43}$$

$H_{\text{medium strong}}$ is responsible for the breaking of the symmetry, e.g. for the mass differences, and H_{SU_3} for the *presence* of the symmetry. ϵ is a parameter which measures the breaking of the symmetry, so from above $\epsilon \sim \Delta m/m \sim 0.15$. Just as in the discussions in Chapter 3 about turning on and off the electromagnetic interactions, we may imagine that it is possible to turn on and off $H_{\text{medium strong}}$ (though it is not!). This may be represented by changing the value of ϵ. In that case, as $\epsilon \to 0$, the supermultiplets become degenerate (if electromagnetism is ignored).

Many physicists feel, however, that it is entirely inappropriate to discuss strong interactions in the language of Hamiltonians. According to this point of view, equation (43) is perhaps a useful way of describing the situation, but no more.

According to yet another school of thought, SU_3 is *intrinsically* broken; that is, there is not a symmetric state of affairs followed by a perturbation which breaks the symmetry, but rather the symmetry when it first appears is already broken.

We can only conclude that at present SU_3 breaking is a major unsolved problem.

c) *Isospin and Strangeness are not Independent*

This may sound a surprising statement, but it is indeed one of the implications of SU_3 that there is a connection between I and S. For, example, if we discover an isospin multiplet of particles, with a particular strangeness, we predict other isospin multiplets with *different* values of S, according to the rules of Figure 6; the values of I must differ by ½ or 1, and of S by 1 or 0, respectively. Arbitary combinations of I and S are *not* allowed. In this sense I and S are not independent. (An interesting historical parallel is space and time; until Einstein, these also were thought to be independent.)

d) *SU_3 does not predict what supermultiplets exist*

This is an important point. Once we have identified some members of a supermultiplet, SU_3 demands that the rest of the members exist; but it does not dictate what supermultiplets should exist in the first place. For instance, SU_3 by itself does not tell us that octets should occur in nature. Any theory which does that must lie beyond SU_3. It is interesting that the only supermultiplets found in nature are singlets, octets and decuplets. Why should this be? This question is taken up in the next chapter.

THE CONSERVED VECTOR CURRENT THEORY

FURTHER READING

CVC theory

Feynman, R. P. and M. Gell-Mann, "Theory of the Fermi interaction", *Physical Review* **109** 193 (1958).

Wu, C. S., "The Universal Fermi Interaction and the Conserved Vector Current in Beta Decay", *Reviews of Modern Physics* **36**, 618 (1964).

Unitary symmetry

Barnes, V. E. et al., "Observation of a Hyperon with Strangeness Minus Three", *Physical Review Letters* **12**, 204 (1964). Reprinted in M. Gell-Mann and Y. Ne'eman, op.cit.

Chew, G. F., M. Gell-Mann and A. H. Rosenfeld, "Strong Interacting Particles", *Scientific American*, February, 1964.

Gell-Mann, M., "The Eightfold Way: A Theory of Strong Interaction Symmetry", Caltech report (1961), unpublished. Reprinted in M. Gell-Mann and Y. Ne'eman, op.cit.

Gell-Mann, M. and Y. Ne'eman, *The Eightfold Way*, Benjamin, 1964.

Lichtenberg, D. B. *Unitary Symmetry and Elementary Particles*, Academic Press, 1970.

Matthews, P. T., *Introduction to Quantum Mechanics*, second edition, McGraw-Hill, 1968, Chapter 14.

Matthews, P. T., "Unitary Symmetry", in *High Energy Physics* vol.I (E. H. S. Burhop, ed.), Academic Press, 1967.

CHAPTER 11

Unitary Symmetry and the Quark Model

11.1 THE MARRIAGE OF ISOSPIN AND STRANGENESS

UNITARY symmetry, SU_3, was derived in the last chapter from weak interaction selection rules. As promised, I shall now show how it can be arrived at purely from a consideration of strong interactions. In this context the problem is simply how to *combine* isospin and strangeness, the two quantum numbers conserved by the strong interaction. By combining them into a "higher dimensional" quantum number (in the sense that isospin is a three dimensional quantum number, whereas charge is only one dimensional), restrictions will automatically be imposed on the allowed combinations of I and S that particles may have, and this will be one way of bringing order into the chaos of the population explosion among particles.

The question is, how this should be done. Perhaps the easiest way to answer is to follow the historical development of the idea that all particles are made out of a basic set of particles. In the case of isospin, if all particles are constructed from one multiplet, this would most simply be a fermion multiplet with $I = \frac{1}{2}$. Any value of I may be constructed by combining enough $I = \frac{1}{2}$ states; and of course any basic particle must be a fermion, since bosons can be made from fermions, but not vice versa. The obvious candidate is the nucleon. As an example the pion may be considered as a nucleon-antineutron bound state with

$$\pi^+ = p\bar{n}$$

$$\pi^0 = \sqrt{\tfrac{1}{2}}(p\bar{p} - n\bar{n})$$

$$\pi^- = n\bar{p}$$

as in (2-39). This model was first proposed in 1949 by Fermi and Yang. There are several theoretical problems connected with it, but for the purpose of instruction it is best to suspend criticism and carry on to ask the question: if non-strange particles are made out of p and n, what are strange particles made out of? Put another way, how may the Fermi–Yang theory be extended to include strange particles?

This question was taken up by Sakata who in 1956 proposed that all hadrons are composites of the three basic states (p, n, Λ), shown in Figure 1. This is

UNITARY SYMMETRY AND THE QUARK MODEL

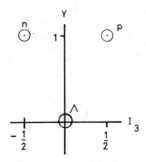

Figure 11.1 p, n and Λ in the Sakata model

simply the Fermi–Yang theory with Λ added. Examples of the way other particles would be constructed are $\pi^+ = p\bar{n}$, $K^+ = p\bar{\Lambda}$, $\Sigma^+ = p\Lambda\bar{n}$, $\Xi^- = \Lambda\Lambda\bar{p}$, $K^0 = n\bar{\Lambda}$, etc. The Sakata model marked the birth of SU_3, for it was pointed out by Ikeda, Ogawa and Ohnuki that just as (p, n) with $I = \frac{1}{2}$ form the lowest "nontrivial" representation of the isospin group SU_2, so, according to the Sakata model, (p, n, Λ) form the lowest non-trivial representation of SU_3. (The "trivial" representation in both cases is the singlet: $I = 0$ in SU_2, $I = Y = 0$ in SU_3.)

Having identified the basic particles, the next question is, what is the rule for combining them into composite states? For isospin, the rule is the vector addition rule. We want the analogous rule for SU_3. The easiest way to find it is to state the vector addition rule in diagrammatic terms, which are then simply extendable to SU_3.

11.2 BUILDING UP ISOSPIN MULTIPLETS

As an example of the vector addition rule, a particle which is a composite of particles with $I = 1$ and $I = \frac{1}{2}$, itself has $I = \frac{3}{2}$ or $I = \frac{1}{2}$: we write

$$(I = 1) \times (I = \tfrac{1}{2}) = (I = \tfrac{3}{2}) + (I = \tfrac{1}{2}). \tag{11-1}$$

Now a particle with $I = 1$ has three states $I_3 = 1, 0, -1$; and particles with $I = \frac{1}{2}$ and $I = \frac{3}{2}$ have respectively two and four states, so we may write instead of (1)

$$3 \times 2 = 4 + 2 \tag{11-2}$$

where the numbers stand for the number of states; and since $3 \times 2 = 4 + 2$, the number of states is the same on both sides of the equation. Writing the states out explicitly, we may represent the left hand side of (2) by

$$\underset{-1\ \ \ 0\ \ \ 1}{\bullet\!\!-\!\!\bullet\!\!-\!\!\bullet} \quad \times \quad \underset{-1\ \ \ 0\ \ \ 1}{-\!\!\bullet\!\!-\!\!\bullet\!\!-} \tag{11-3}$$

where the dots denote the values of I_3. This is the product of three states with two states, which may be calculated as follows. Since the values of I_3 simply add, $I_3 = 1$ combined with $I_3 = \pm \tfrac{1}{2}$ gives $I_3 = \tfrac{3}{2}, \tfrac{1}{2}$. $I_3 = 0$ combined with $I_3 = \pm \tfrac{1}{2}$ gives $I_3 = \pm \tfrac{1}{2}$, and $I_3 = -1$ combined with $I_3 = \pm \tfrac{1}{2}$ gives $I_3 = -\tfrac{3}{2}, -\tfrac{1}{2}$. So we have

$$\bullet\!\!-\!\!\bullet\!\!-\!\!\bullet \quad \times \quad \bullet\!\!-\!\!\bullet\!\!-\!\!\bullet \quad = \quad \bullet\!\!:\!\!\bullet\!\!:\!\!\bullet$$
$$-1 \quad 0 \quad 1 \qquad -1 \quad 0 \quad 1 \qquad -\tfrac{3}{2}\; -\tfrac{1}{2}\; \tfrac{1}{2}\; \tfrac{3}{2}$$
(11-4)

In other words, to get the rhs of (4), superimpose the *centre* $(I_3 = 0)$ of the $I = \tfrac{1}{2}$ diagram on each of the three states $I_3 = 1, 0, -1$ of the $I = 1$ diagram, and mark the new positions of the states $I_3 = \pm \tfrac{1}{2}$. This gives the 6 dots in (4), which are then split up into states with a unique isospin, so that we have

$$\bullet\!\!-\!\!\bullet\!\!-\!\!\bullet \quad \times \quad \bullet\!\!-\!\!\bullet\!\!-\!\!\bullet \quad = \quad \bullet\!\!-\!\!\bullet\!\!-\!\!\bullet\!\!-\!\!\bullet \quad +$$
$$-1 \quad 0 \quad 1 \qquad -1 \quad 0 \quad 1 \qquad -\tfrac{3}{2}\; -\tfrac{1}{2}\; \tfrac{1}{2}\; \tfrac{3}{2}$$

$$\bullet\!\!-\!\!\bullet\!\!-\!\!\bullet$$
$$-1 \quad 0 \quad 1$$
(11-5)

The manipulations (4) and (5) are summed up in the *rule*

To obtain the product of two isospins, superimpose the centre of the diagram for one isospin on each of the states in turn of the diagram for the other isospin, and mark the resulting positions of the states of the first isospin. Split up the resulting diagram into diagrams with a unique value of I. (11-6)

11.3 BUILDING UP SU$_3$ MULTIPLETS

It is now easy to generalise the rule (6) to unitary spin, but before doing so, it is best to loosen our ties with the Sakata model. Let us assume that there are 3 basic particles, but not that they are p, n and Λ. In fact let us abandon the historical approach and take up the suggestion of Gell-Mann in 1964 that the three basic particles, which he called *quarks*, do not have the same hypercharge Y as p, n and Λ, but are as shown in Figure 2.[†] (u and d stand for isospin up and down, s for singlet.) The reason for choosing fractional values of Y will become clear later. The antiquarks $\bar{u}$, $\bar{d}$ and $\bar{s}$ have opposite values of I_3 and Y, and appear also in Figure 2. The quark and antiquark supermultiplets are denoted **3** and **3*** respectively.

[†] Quarks were postulated simultaneously by Zweig, who called them "aces". The word quark comes from James Joyce's *Finnegans Wake* ("Three quarks for Muster Mark!").

UNITARY SYMMETRY AND THE QUARK MODEL   193

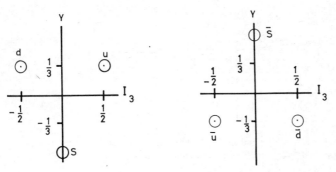

Figure 11.2   The supermultiplets 3 and 3* — quarks and antiquarks

Let us now see what states result by combining quarks with antiquarks. *To do this we use the rule (6), reading "unitary spin" instead of "isospin".* This means we superimpose the centre of the diagram 3* on each of the three states of 3, and mark the positions of the 9 resulting states. The resulting pattern is the one in Figure 3, which contains 3 states at the origin. This may be split up into a supermultiplet of **8** and one of **1**, as in Figure 4, so that we have

$$3 \times 3^* = 8 + 1 \tag{11-7}$$

This is the SU_3 (unitary spin) analogue of equation (2). It is obtained from rule (6), suitably reworded:

To obtain the product of two unitary spins, superimpose the centre of the diagram for one unitary spin on each of the states in turn of the diagram for the other unitary spin, and mark the resulting positions of the states of the first unitary spin. Split up the resulting diagram into diagrams with a unique value of unitary spin (11-8)

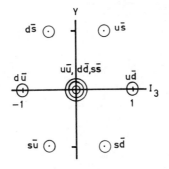

Figure 11.3

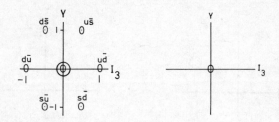

Figure 11.4 The octet 8 and singlet 1. The octet states with $I_3 = Y = 0$ are $(2s\bar{s} - u\bar{u} - d\bar{d})/\sqrt{6}$, and $(u\bar{u} - d\bar{d})/\sqrt{2}$ and the singlet state $(u\bar{u} + d\bar{d} + s\bar{s})/\sqrt{3}$

The octet in (7) is the same one as in Figures 10-7 and 10-8. In words, equation (7) tells us that an arbitary compound state of a quark and an antiquark belongs to either an octet or a singlet of SU_3.

Let us now consider what happens when we combine quarks with quarks. We proceed in the same way as before, using rule (8) and the quark diagram for 3 in Figure 2. It is easily seen that

$$3 \times 3 = 6 + 3^* \tag{11-9}$$

where the supermultiplet **6** — the sextet — is as shown in Figure 5.

EXERCISE Prove equation (9).

We now add another quark to the two quarks. That is, we want to calculate $3 \times 3 \times 3$. Using (9) we have

$$3 \times 3 \times 3 = (6 + 3^*) \times 3$$

$$= (6 \times 3) + (3^* \times 3). \tag{11-10}$$

By rule (8) we get

$$6 \times 3 = 10 + 8 \tag{11-11}$$

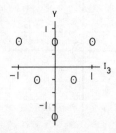

Figure 11.5 The sextet 6

UNITARY SYMMETRY AND THE QUARK MODEL

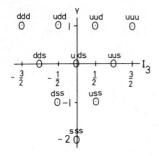

Figure 11.6   The decuplet **10**

where **10** is the decuplet shown in Figure 6, and **8** is the octet in Figure 4. For convenience, the u, d and s content of each state in **10** has been shown. **10** will be recognised as the same as the decuplet in Figure 10-9.

EXERCISE   Prove equation (11).

3* × **3** on the rhs of (10) we have already found in (7), so by combining (7), (10) and (11) we now have†

$$3 \times 3 \times 3 = 10 + 8 + 8 + 1 \tag{11-12}$$

In words, (12) says that a state of three quarks belongs either to a decuplet, an octet or a singlet.

11.4 THE QUARK MODEL

We saw in the last chapter that both the $\tfrac{1}{2}^+$ baryons and the 0^- mesons belong to octets, and the $\tfrac{3}{2}^+$ baryons to a decuplet. On the other hand we have seen above that $q\bar{q}$ belong to an octet or a singlet, and qqq to a decuplet, octet, or singlet. Let us now suppose that

all mesons are bound states of $(q\bar{q})$

all baryons are bound states of (qqq). \hfill (11-13)

The quarks must be fermions, and the simplest spin they may have is $\tfrac{1}{2}$. They must also have $B = \tfrac{1}{3}$, so that the baryons have $B = 1$. The values of Q then follow from the Gell-Mann–Nishijima relation

---

† Let us note in passing that equations (7), (10)–(12) all "agree" with simple arithmetic: $3 \times 3 = 9 = 6 + 3; 6 \times 3 = 18 = 10 + 8; 3 \times 3 \times 3 = 27 = 10 + 8 + 8 + 1$.

$$Q = I_3 + \frac{B+S}{2}$$

and have the remarkable values ⅔ and −⅓. The quark quantum numbers are summarised in Table 1.

Referring to (13), the lowest mass and lowest spin particles will be those with zero angular momentum between the quarks, $L = 0$. Higher spin particles are simply the result of higher L. This "orbital excitation" will automatically give an increased mass.

To examine the model in more detail, it is best to treat mesons and baryons separately.

Mesons

If mesons are bound $(q\bar{q})$ states then according to (7) they must all belong to unitary octets or singlets. That is, there should be no mesons with $I = \frac{3}{2}$ (e.g. $K^+\pi^+$ or $K^-\pi^-$ resonances), with $I = 2$ (e.g. $\pi^+\pi^+$ or $\pi^-\pi^-$ resonances), or with $Y = S = \pm 2$ (K^+K^+ or K^-K^- resonances), since these would not fit into an octet. This is indeed what is found; despite considerable search, no resonances have ever been discovered in systems such as the above ones. (See Section 7.11.)

The $q\bar{q}$ system has spin $S = 1$ or 0, and since q and $\bar{q}$ have opposite parity, the parity $= (-1)^{L+1}$ where L is the orbital angular momentum. Hence the lowest lying states (which have the lowest value of L) are

$$L = 0, \quad J = 0, 1 \quad \begin{array}{l} \to 0^- \\ 1^- \end{array} \quad \left. \begin{array}{l} {}^1S_0 \\ {}^3S_1 \end{array} \right\} \; 8, \; 1$$

$$L = 1, \quad J = 0, 1, 2 \to \begin{array}{l} 0^+ \\ 1^+ \\ 1^+ \\ 2^+ \end{array} \quad \left. \begin{array}{l} {}^3P_0 \\ {}^3P_1 \\ {}^1P_1 \\ {}^3P_2 \end{array} \right\} \; 8, \; 1 \qquad (11\text{-}14)$$

where the notation $^{2S+1}L_J$ has been used. The quark model then predicts the existence of meson octets and singlets with the above values of spin and parity, and a general increase in mass with L, and therefore with spin. The known meson

Table 11.1
The quantum numbers of the quarks u, d and s

Quark	I	I_3	S	B	Q
u	1/2	1/2	0	1/3	2/3
d	1/2	−1/2	0	1/3	−1/3
s	0	0	−1	1/3	−1/3

UNITARY SYMMETRY AND THE QUARK MODEL   197

states have already appeared in Figure 7-4, where several octets and singlets have been identified. Some particles have still to be discovered for the multiplets to be complete. It may be seen that in general terms the pattern of states indicated by (14) above corresponds quite well with the existing states. The lowest mass octets are the 0^- and 1^-, and there are also singlets of these quantum numbers, $\eta'(958)$ (which may however have $J^P = 2^-$) and $\varphi(1019)$ or $\omega(784)$; which of ω and φ is the SU_3 singlet with $I = Y = 0$ and which is the $I = Y = 0$ member of the octet is not easy to decide, since there is no great difference in mass, as there is between $\eta(549)$ and $\eta'(958)$. This question is discussed in section 11.5D below. There is a partially complete 0^+ octet and singlet, and the makings of two 1^+ octets, as suggested by (14). Finally the complete 2^+ octet and singlet have been found. This information appears in Table 2.

Baryons

If baryons are bound qqq states, then according to (12) they must all belong either to decuplets, octets or singlets. This is consistent with observation; no baryon states have been found with $Y = 2$ (K^+p resonances) or with $Y = -1, I = \tfrac{3}{2}$ ($\pi^-\Xi^-$ resonances) — see section 7.11. If such states were found they would have to be placed in other multiplets of SU_3, for instance **27** or **10***, which would correspond to something other than qqq; for example $qqqq\bar{q}$.

If the total angular momentum of the three quarks is L, then we may put

$$L = l_1 + l_2$$

where l_1 is the relative angular momentum of two of the quarks in their centre of mass system, and l_2 the angular momentum of the third quark relative to the centre of mass of the first two. According to the orbital excitation hypothesis, the lowest lying baryons will have $L = l_1 = l_2 = 0$ and parity $(-1)^L = +1$ (taking

Table 11.2
Meson octets and singlets from the quark model with $L = 0, 1$. ? denotes that the state has not been established

$2S+1_{L_J}$	J^P	Octet and singlet states			
		$I = 1, Y = 0$	$I = \tfrac{1}{2}, Y = \pm 1$	$I = 0, Y = 0$	$I = 0, Y = 0$
$^1S_0$	0^-	$\pi(135)$	$K(498)$	$\eta(549)$	$\eta'(958)$
$^3S_1$	1^-	$\rho(765)$	$K^*(891)$	$\omega(784)$	$\varphi(1019)$
$^3P_0$	0^+	$\delta(962)$	?	$\epsilon(700-1000)$	$S^*(1070)$
$^3P_1$	1^+	$A1(1070)$	$K_A(1240)$	$D(1285)$	?
$^1P_1$	1^+	$B(1235)$	$K_A(1320)$	?	?
$^3P_2$	2^+	$A2(1310)$	$K_N(1420)$	$f(1260)$	$f'(1514)$

quark parity as positive). Since by the vector addition rule the spin of 3 spin ½ particles is $S = \frac{3}{2}$ or $\frac{1}{2}$, then these baryons have total angular momentum given by

$$J^P = \tfrac{3}{2}^+, \tfrac{1}{2}^+.$$

The nucleon octet and the $\Delta(1236)$ decuplet have these values of J^P, as noted in Table 3.

The next lowest mass baryons will have $L = 1$, i.e. either $l_1 = 1, l_2 = 0$ or $l_1 = 0$, $l_2 = 1$, and parity $(-1)^L = -1$. By the vector addition rule the allowed values of J^P will then be

$$J^P = \tfrac{1}{2}^-, \tfrac{3}{2}^-, \tfrac{5}{2}^-.$$

By inspection of the baryon resonances appearing in Figure 7-2, we may identify several supermultiplets with these values of J^P. They are noted in Table 3.

11.5 MASS FORMULAE

The fact that SU_3 is a broken symmetry makes it rather difficult to deal with; this problem is illustrated by the masses of particles in octets and decuplets. If the symmetry were exact, the particles would be degenerate. At the time that SU_3 was proposed, however, a "mass formula" was also proposed by Gell-Mann and Okubo. According to this formula, the masses of particles in a supermultiplet are *not random*, but obey a fairly simple relation, corresponding to the fact that the symmetry is broken in a fairly simple way. The original derivation of the Gell-Mann–Okubo mass formula was a group theoretic one, and is beyond the scope of this book. It turns out, however, that it can be derived from the quark model by assuming that u and d have the same mass (isospin symmetry), but that s has

Table 11.3
Baryon supermultiplets from the quark model with $L = 0, 1$. * denotes that the state is not firmly established

	J^P	Supermultiplet			
$L=0$	$1/2^+$	8: $N(939)$,	$\Lambda(1115)$,	$\Sigma(1192)$,	$\Xi(1315)$
	$3/2^+$	10: $\Delta(1236)$,	$\Sigma(1385)$,	$\Xi(1530)$,	$\Omega(1672)$
$L=1$	$1/2^-$	1: $\Lambda(1405)$			
	$3/2^-$	1: $\Lambda(1520)$			
	$1/2^-$	8: $N(1535)$,	$\Lambda(1670)$,	$\Sigma(1750)$,	$\Xi(1820)^*$
	$3/2^-$	8: $N(1520)$,	$\Lambda(1690)$,	$\Sigma(1670)^*$,	$\Xi(1820)^*$
	$5/2^-$	8: $N(1670)$,	$\Lambda(1830)$,	$\Sigma(1765)$,	$\Xi(1940)^*$

UNITARY SYMMETRY AND THE QUARK MODEL

a different mass. Knowing the quark content of the particles, a relation between the masses follows immediately. This section is devoted to deriving these formulae.

A 0^- meson octet

By comparing Figures 4 and 10-8, we make the following identifications

$$K^+ = u\bar{s}, \quad K^0 = d\bar{s} \tag{11-15}$$

$$\pi^+ = -u\bar{d}, \quad \pi^0 = (u\bar{u}, d\bar{d}, s\bar{s}), \quad \pi^- = d\bar{u} \tag{11-16}$$

$$\eta = (u\bar{u}, d\bar{d}, s\bar{s}) \tag{11-17}$$

$$\eta' = (u\bar{u}, d\bar{d}, s\bar{s}) = SU_3 \text{ singlet} \tag{11-18}$$

$$\overline{K^0} = -s\bar{d}, \quad K^- = s\bar{u} \tag{11-19}$$

where π^0, η and η' are made up of combinations of $u\bar{u}$, $d\bar{d}$ and $s\bar{s}$ yet to be found. Let us first find π^0. We do this by utilising the operators I_+ and I_- defined in Chapter 3. From equation (3-36), we have for example

$$I_-\pi^+ = \sqrt{2}\pi^0 \tag{11-20}$$

Now look at (16). The doublets (u, d) and $(-\bar{d}, \bar{u})$ behave under I_- just as (p, n) and $(-\bar{n}, \bar{p})$, so that, in analogy with (3-34) and (3-35)

$$I_-u = d, \quad I_-\bar{d} = -\bar{u}$$

so that

$$I_-(-u\bar{d}) = -(I_-u)\bar{d} - u(I_-\bar{d})$$

$$= -d\bar{d} + u\bar{u}. \tag{11-21}$$

Comparing this with (20), since $\pi^+ = -u\bar{d}$ from (16), we have

$$\pi^0 = \sqrt{\tfrac{1}{2}}(u\bar{u} - d\bar{d}). \tag{11-22}$$

The other unknown states are η and η'. η' is easy to write down, since the fact that it is an SU_3 singlet means that it is an I-spin, a U-spin and a V-spin singlet, and therefore should be a symmetric combination of $u\bar{u}$, $d\bar{d}$ and $s\bar{s}$. This is

$$\eta' = \sqrt{\tfrac{1}{3}}(u\bar{u} + d\bar{d} + s\bar{s}). \tag{11-23}$$

η must now be orthogonal to both π^0 and η', which requires

$$\eta = \sqrt{1/6}(2s\bar{s} - u\bar{u} - d\bar{d}). \tag{11-24}$$

EXERCISES 1) Prove that π^0, η and η' are mutually orthogonal.

2) By using (3-37) for U-spin, rather than I-spin,

$$U_+(U, U_3) = \sqrt{(U - U_3)(U + U_3 + 1)}(U, U_3 + 1)$$

show that the U-spin triplet in the octet is

$(K^0, \tfrac{1}{2}(\pi^0 + \sqrt{3}\eta), \overline{K^0})$.

3) Show that η' has $I = U = V = 0$.

Now denoting

$$m_u = m_d = m_{\bar{u}} = m_{\bar{d}} = m_1$$

$$m_s = m_{\bar{s}} = m_2 \tag{11-25}$$

and supposing that the quarks and antiquarks are bound with a binding energy b which is the same for each particle, the mass of K^0, for example, may be calculated as follows. Using (15), the mass is simply the expectation value of the energy operator (at zero momentum):

$$m_{K^0} = \int (\psi_{\bar{s}}\psi_d)^* H \psi_{\bar{s}}\psi_d \, d\tau - b$$

$$= \int (\psi_{\bar{s}}\psi_d)^* (H_{\bar{s}} + H_d)(\psi_{\bar{s}}\psi_d) \, d\tau - b$$

$$= \int (\psi_{\bar{s}}^* H_{\bar{s}} \psi_{\bar{s}})(\psi_d^* \psi_d) \, d\tau + \int (\psi_{\bar{s}}^* \psi_{\bar{s}})(\psi_d^* H_d \psi_d) \, d\tau - b$$

$$= m_{\bar{s}} + m_d - b$$

$$= m_1 + m_2 - b$$

where (25) has been used. The masses of the other particles are calculated similarly, using the facts that the wave functions are normalised to $\int \psi_a^* \psi_b \, d\tau = 0$ if $a \neq b$. In this way we get

$$m_K = m_1 + m_2 - b$$

UNITARY SYMMETRY AND THE QUARK MODEL

$m_\pi = 2m_1 - b$

$m_\eta = \tfrac{2}{3}m_1 + \tfrac{4}{3}m_2 - b$

$m_{\bar{K}} = m_1 + m_2 - b.$ (11-26)

On elimination of m_1, m_2 and b this gives

$$\frac{m_K + m_{\bar{K}}}{2} = \frac{3m_\eta + m_\pi}{4},$$ (11-27)

which, since $m_K = m_{\bar{K}}$, simplifies to

$m_K = \tfrac{1}{4}(3m_\eta + m_\pi).$ (11-28)

The lhs of (28) is 498 MeV, and the rhs = 446 MeV. The discrepancy is about 11%, and may be improved if (mass)^2 is substituted for (mass) in (28). We then have

$$m_K^2 = \frac{3m_\eta^2 + m_\pi^2}{4}$$ (11-29)

whose lhs = 0.248 (GeV)^2 and rhs = 0.230 (GeV)^2; they agree to within 8%. It is now generally accepted that for pseudoscalar mesons the mass formula is a (mass)^2 formula, though the reason for this is not fully understood.

EXERCISES   1) Prove equations (26).

2) Adopt the point of view that the formula (28) should really be a formula for the *energies* E_i of the particles i, where $E_i^2 = m_i^2 + p^2$ is the particle has momentum p. Assuming that in the limit of perfect symmetry, the 0^- mesons have *zero* mass, prove (29).

B   Baryon 3/2$^+$ decuplet

This mass formula is almost trivial to derive. Simply compare Figures 10-8 and 11-6, and use (25) to give

$m_\Delta = 3m_1 - B_0$

$m_{Y_1^*} = 2m_1 + m_2 - B_0$

$m_{\Xi^*} = m_1 + 2m_2 - B_0$

$$m_\Omega = 3m_2 - B_0 \tag{11-30}$$

where B_0 is the 3-quark binding energy. The equal spacing rule (10-42) follows immediately.

C Baryon ½⁺ octet

The quark content of the members of the $½^+$ octet is exactly the same as that of the members of the $\tfrac{3}{2}^+$ decuplet with the same I_3 and Y. Comparing Figures 10-7 and 10-9 and using (30) then gives

$$\Lambda \to N: m_N = 3m_1 - B_1$$

$$Y_1 \to \begin{cases} \Sigma: m_\Sigma = 2m_1 + m_2 - B_1 \\ \Lambda: m_\Lambda = 2m_1 + m_2 - B_1 \end{cases}$$

$$\Xi^* \to \Xi: m_\Xi = m_1 + 2m_2 - B_1. \tag{11-31}$$

These imply that

$$\frac{m_N + m_\Xi}{2} = \frac{3m_\Lambda + m_\Sigma}{4} \tag{11-32}$$

This formula is extremely well satisfied; taking the neutral particles, lhs = 1127 MeV, rhs = 1135 MeV, a discrepancy of less than 1%. (31) also implies the bad relation $m_\Sigma = m_\Lambda$, but this is not given by the more general Gell-Mann–Okubo mass formula (see later), and so may be overlooked as being a peculiarity of the quark model.

D Meson 1⁻ singlet and octet

A straight substitution from (28) $K \to K^*$, $\eta \to \omega$ and $\pi \to \rho$ would give

$$m_{K^*} = \frac{3m_\omega + m_\rho}{4}$$

whose lhs = 892 MeV and rhs = 778 MeV, a discrepancy of about 14%. This is rather large, and suggests that we should not make the above substitutions; but this in turn would imply that K^*, ρ, ω and $\overline{K}^*$ do not form a true octet, for if they did, they would be equivalent to K, π, η and $\overline{K}$. How could this be? The clue comes from the presence of the φ meson, close in mass to the ω meson

$$\varphi(1019): I = 0, S = 0$$

UNITARY SYMMETRY AND THE QUARK MODEL

$\omega(784)$: $I = 0, S = 0$.

Now imagine a general situation in which there are two particles with the same J^P and B, with $I = 0$ and $S = 0$, but with different masses, such that one belongs to an octet, and the other is a singlet. In other words, all the quantum numbers are identical *except* for the unitary spin; and the masses are different. If unitary symmetry were exact there would be no complication, but the fact is that it is not exact. This means that the symmetry breaking part of the strong interactions may *change* the unitary spin of a state, though it cannot change the isospin or strangeness (which strong interactions conserve exactly), so it may happen that

$$(I = 0, Y = 0 \text{ octet}) \xleftarrow{\text{SU}_3 \text{ breaking} \atop \text{strong interactions}} (I = 0, Y = 0 \text{ singlet})$$

and the particles *mix*.

Let us apply this idea to ω and φ. Suppose that in a world which is exactly SU_3 symmetric, the particles belonging to the octet and singlet are respectively φ_8 and φ_1. Now "turn on" the symmetry breaking forces, so that these particles mix to produce the real, observed ω and φ. This mixing is given by

$$\varphi = \varphi_8 \cos\theta + \varphi_1 \sin\theta$$
$$\omega = -\varphi_8 \sin\theta + \varphi_1 \cos\theta \qquad (11\text{-}33)$$

so that the physical particles ω and φ do not belong to a unique supermultiplet of SU_3. θ is called the *mixing angle*.

EXERCISE Prove that if φ_8 and φ_1 are orthonormal, then so are ω and φ.

We now make a specific *assumption*: that φ is made out of s and $\bar{s}$ quarks only, and that ω is made out of $u, d, \bar{u}$ and $\bar{d}$ quarks only:

$$\varphi = s\bar{s}$$
$$\omega = \sqrt{\tfrac{1}{2}}(d\bar{d} + u\bar{u}). \qquad (11\text{-}34)$$

φ_8 is analogous to η, and φ_1 to η', so from (23) and (24)

$$\varphi_8 = \sqrt{\tfrac{1}{6}}(2s\bar{s} - d\bar{d} - u\bar{u})$$
$$\varphi_1 = \sqrt{\tfrac{1}{3}}(s\bar{s} + d\bar{d} + u\bar{u}). \qquad (11\text{-}35)$$

From (34) and (35)

$$\varphi = \sqrt{2/3}\,\varphi_8 + \sqrt{1/3}\,\varphi_1$$
$$\omega = -\sqrt{1/3}\,\varphi_8 + \sqrt{2/3}\,\varphi_1, \qquad (11\text{-}36)$$

which show that φ spends, as it were, $2/3$ of its life belonging to an octet, and $1/3$ as a singlet; and ω the other way round. The mixing angle in this case is given by $\sin\theta = \sqrt{1/3}$, $\theta \approx 35°$. (34) implies that

$$\varphi_8 = \sqrt{2/3}\,\varphi - \sqrt{1/3}\,\omega$$
$$\varphi_1 = \sqrt{1/3}\,\varphi + \sqrt{2/3}\,\omega. \qquad (11\text{-}37)$$

The particles making the true octet of 1^- mesons are ρ, φ_8, K^*, and they will obey the mass formula of any octet (cf (28))

$$3m_{\varphi_8} + m_\rho = 4m_{K^*}.$$

Finally, using (37) we get

$$2m_\varphi + m_\omega + m_\rho = 4m_{K^*}. \qquad (11\text{-}38)$$

The lhs of (38) is 3588 MeV, and the rhs 3568 MeV; the formula holds to a good accuracy.

E Gell-Mann–Okubo mass formula

The octet and decuplet mass formulae above are examples of the general Gell-Mann–Okubo mass formula, which is

$$m = m_0 + m_1 S + m_2[\tfrac{1}{4}S^2 - I(I+1)] \qquad (11\text{-}39)$$

for a particle of strangeness S and isospin I. With the qualifications that m^2 rather than m should be used for the 0^- mesons, and that there is mixing for the 1^- mesons, the reader may verify that (39) implies the formulae quoted above.

11.6 MASS FORMULAE FOR ELECTROMAGNETIC MASS DIFFERENCES

The mass formulae discussed so far relate the masses of different isospin multiplets in the same unitary spin family, and were derived by assuming a specific model for the way in which unitary spin is not conserved. This model is one where the u and d quarks have mass m_1 and the s quark mass m_2 ($\neq m_1$), so isospin symmetry holds. To work out how the particle masses change when we turn on the electromagnetic interactions, we need to assume something about how the electromagnetic interactions behave with respect to unitary spin.

Consider for example the baryon ½⁺ octet. Comparing Figures 10-6 and 10-7, we may write out the U-spin multiplets:

$U = ½ \quad (p, \Sigma^+)$

$U = 1 \quad (n, a\Sigma^0 + \beta\Lambda, \Xi^0)$

$U = 0 \quad (\beta\Sigma^0 - a\Lambda)$

$U = ½ \quad (\Sigma^-, \Xi^-)$

where a and β (with $|a|^2 + |\beta|^2 = 1$) are coefficients which determine the proportions of Σ^0 and Λ in the U-spin triplet and singlet. (Cf. Exercise (2), p.200). The point to notice is that *all particles in the same U-spin multiplet have the same charge*; so we may make an educated guess that *electromagnetic interactions conserve U-spin*, that is, they affect all members of a U-spin multiplet in the same way. Letting particle symbols stand for masses, if the masses before the electromagnetic interactions are turned on are N, Σ, Λ and Ξ, then the presence of the electromagnetic forces introduces the common shift

$p = N + \delta_0$

$\Sigma^+ = \Sigma + \delta_0$ \hfill (11-40)

to the members of the U-spin doublet. Similarly

$n = N + \delta_1$

$\Xi^0 = \Xi + \delta_1$

$\Sigma^- = \Sigma + \delta_2$

$\Xi^- = \Xi + \delta_2.$ \hfill (11-41)

Equations (40) and (41) imply that

$\Xi^- - \Xi^0 = (\Sigma^- - \Sigma^+) + (p - n)$ \hfill (11-42)

whose lhs = 6.6 ± 0.9 MeV, and rhs = (8.0 − 1.3) ± 0.2 = 6.7 ± 0.2 MeV − a remarkable agreement. (42) is called the Coleman–Glashow formula.

11.7 THE SIGNIFICANCE OF THE QUARK MODEL – DO QUARKS EXIST?

The main attraction of the quark model is that it offers an explanation of why the only supermultiplets observed in nature are singlets, octets and decuplets. In addition, the "orbital excitation" model explains the spins and parities of the lowest lying states. In general terms it explains why particles and resonances are not composite states of each other (see Section 7.12); they are all composites of quarks and antiquarks with different *excitation* energies. The mass formulae, which agree so well with experiment, have a simple origin if quarks exist; and it may also be shown from the quark model that the ratio of the neutron and proton magnetic moments is

$$\frac{\mu_n}{\mu_p} = -\frac{2}{3}$$

This is in excellent agreement with the experimental value $-1.91/2.79 = -0.68$.

All this is to the credit of the quark model. On the debit side, quarks have not yet been discovered as individual particles. All the experimental efforts have been directed towards searching for a particle with a charge $Q = \frac{2}{3}$ or $-\frac{1}{3}$. This should be detectable through its weak ionisation, which would be $\frac{4}{9}$ or $\frac{1}{9}$ ($=Q^2$) times the ionisation produced by a singly charged particle. Quarks have been looked for (a) as $q\bar{q}$ pairs produced in high energy accelerators, (b) in cosmic rays, (c) in samples of terrestrial matter.

The results are (a) if $M_q < 5$ GeV, the reaction cross section is $\sigma < 4 \times 10^{-37}$ cm^2 ($Q = \frac{2}{3}$), $\sigma < 3 \times 10^{-39}$ cm^2 ($Q = -\frac{1}{3}$). These figures are from Serpukhov (USSR). Such a small cross section is, of course, incompatible with the very strong interactions quarks must have with hadrons. The implication is that $M_q > 5$ GeV. (b) The flux of quarks in cosmic rays is $<10^{-10}$ quarks/cm^2 ster. sec. (c) From the analysis of samples of graphite, there is less that 1 quark per 2×10^8 nucleons. In this connection it should be borne in mind that at least *one* of the quarks must be stable – fractionally charged particles can decay into one another, but not into integrally charged particles – so any quarks bombarding the earth at any time during the last 10^9 years must still be here.

In view of this, many physicists are now coming round to the view that quarks do not exist – that they are "mathematical entities" only. This makes the success of the quark model more mysterious.

11.8 A NOTE ON ELEMENTARITY

If quarks exist, they are obviously the truly "elementary particles". Even if they do not exist, however, there is a lesson to be learned from the fact that all hadrons belong to singlets, octets or decuplets. To see this, let us suppose that some par-

UNITARY SYMMETRY AND THE QUARK MODEL

ticles are more elementary than others. Which particles would we elevate to the honour of being elementary?

As a first guess, we could suppose that the *stable* particles are truly elementary, and the rest composite. This viewpoint is easy to discredit; for consider the $\frac{3}{2}^+$ decuplet. Δ, Y_1^* and Ξ^* are all unstable, and Ω is stable (with respect to the strong interactions, of course). But the whole idea of unitary symmetry is that particles belonging to the same supermultiplet are states of the same particle, so there can be no fundamental difference between Ω and the other members of the decuplet: *stability is not a criterion of elementarity*.

As a second guess, then, let us suppose that the $\frac{1}{2}^+$ and 0^- octets are elementary and the other supermultiplets composite. According to this view the members of the decuplet would be composite states of the $\frac{1}{2}^+$ and 0^- particles, with $l = 1$; in common parlance, Δ is a "pion-nucleon resonance". A composite of two members of octets would belong to one of the supermultiplets contained in the product 8×8, which is

$$8 \times 8 = 27 + 10 + 10^* + 8 + 8 + 1 \qquad (11\text{-}43)$$

as the hard working reader may verify, where 10^* is 10 with $Y \to -Y$, and 27 is hexagonal in shape like 8, and is got from 8 by adding one more "layer" of states' and by increasing the degeneracy of all the other states by 1. There would then be no more reason for the composite state to belong to 10 than for it to belong to 10^* or 27. But there are no known particles belonging to 10^* or 27. It is most logical to regard the 10 that exists not as the one appearing in (43), but as the one in $3 \times 3 \times 3 = 10 + 8 + 8 + 1$; it is *not* a composite of 0^- mesons and $\frac{1}{2}^+$ baryons.

We are then lead to the view that the $\frac{3}{2}^+$ decuplet as just as elementary as the $\frac{1}{2}^+$ or 0^- octets; and in general that no supermultiplet is more elementary than any other one. This reasoning, based purely on the group SU_3, leads, interestingly, to the same conclusion of "nuclear democracy" as was arrived at in section 7.12.

FURTHER READING

See the Further Reading in Chapter 10, under "Unitary Symmetry", and the following
Berman, S. M., "Elements of SU_3", in *Symmetries in Elementary Particle Physics*, (A. Zichichi, ed.), Academic Press, 1965.
Dalitz, R. H., "Quark Models for the Elementary Particles", in *High Energy Physics* (C. DeWitt and M. Jacob, eds.), Gordon and Breach, 1965.
Dalitz, R. H., "Quarks, the Hadronic Sub-units?", in *Contemporary Physics*, International Atomic Energy Agency, Vienna, 1969.
Kokkedee, J. J. J., *The Quark Model*, Benjamin, 1964.
Lipkin, H. J., *Lie Groups for Pedestrians*, North-Holland, 1965.
Lipkin, H. J., "Particle Physics for Nuclear Physicists" in *Physique Nucléaire/Nuclear Physics* (C. DeWitt and V. Gillet, eds.) Gordon and Breach, 1969.
Morpurgo, G. "A Short Guide to the Quark Model", *Annual Reviews of Nuclear Science* **20**, 105 (1970).

CHAPTER 12

Cabibbo's Theory, Chiral Symmetry and Current Algebra

IN THIS CHAPTER we return to a consideration of the coupling constants G_F, G_{GT} and G_μ given in (10-5) and (10-17) as

$$G_F = (1.415 \pm 0.002) \times 10^{-49} \text{ erg cm}^3$$

$$G_{GT} = (1.67 \pm 0.04) \times 10^{-49} \text{ erg cm}^3$$

$$G_\mu = (1.4350 \pm 0.0003) \times 10^{-49} \text{ erg cm}^3$$

$$g_A = G_{GT}/G_F = 1.18 \pm 0.02. \tag{12-1}$$

The near equality

$$G_F \approx G_\mu \tag{12-2}$$

was what sparked off the conserved vector current theory. But the point was made on page 176 that the CVC theory merely guarantees that

$$G_F^{\text{physical}} = G_F^{\text{bare}};$$

in other words, that G_F is not changed from its "bare" value by the pion cloud. The fact that the equality (1) is not exact is a logically independent circumstance. What the exact relation is between G_F and G_μ was elucidated by Cabibbo and in this chapter we begin by considering Cabibbo's theory. We then go on to consider the question: if the weak vector current is conserved, what about the weak axial vector current? This consideration has given rise to a *chiral symmetry* of the strong interactions. Finally a very brief outline is given of Gell-Mann's theory of current algebra, which is an attempt to tackle the problem of broken symmetry. Using current algebra and ideas from chiral symmetry, Adler and Weisberger, in celebrated papers, estimated g_A above, and obtained values very close to the experimental one.

12.1 CABIBBO'S THEORY OF UNIVERSITALITY

The question at hand is the relation between G_F and G_μ. Their closeness beguiled physicists for some years into believing in a form of "simple universality"

Simple universality: $G_F = G_\mu$,　　　　　　　　　　(12-3)

according to which muon decay and Fermi transitions in semi-leptonic $\Delta S = 0$ decays are of the same strength. This simple universality, however, is wrong, for from (10-18)

$$\frac{G_\mu - G_F}{G_F} = 2.8\% \tag{12-4}$$

What, then, is the relation, if any, between G_μ and G_F?

Cabibbo pointed out that, because of SU_3, we cannot consider $\Delta S = 0$ decays without also considering $\Delta S = 1$ decays. (In fact, it was the consideration of both types of decay that lead to SU_3 in the first place.) This means that a universality scheme should also feature the Fermi constant for $\Delta S = 1$ decays, which I denote $G_F{'}$. The situation may be pictured as in Figure 1. G_μ describes the strength of μ decay. G_F denotes the strength of Fermi transitions in *all* $\Delta S = 0$, $|\Delta I_3| = 1$ decays, e.g.

$$G_F: n \to p + e^- + \bar{\nu}_e$$

$$\pi^- \to \pi^0 + e^- + \bar{\nu}_e$$

$$\Sigma^+ \to \Sigma^0 + e^+ + \nu_e, \tag{12-5}$$

and $G_F{'}$ denotes the strength of Fermi transitions in $|\Delta S| = 1$, $|\Delta I_3| = \frac{1}{2}$ ($\Delta S = \Delta Q$) decays, e.g.

$$G_F': \Lambda \to p + e^- + \bar{\nu}_e$$

$$\Sigma^- \to n + \mu^- + \bar{\nu}_\mu$$

$$K^+ \to \pi^0 + e^+ + \nu_e$$

$$\Xi^- \to \Lambda + e^- + \bar{\nu}_e. \tag{12-6}$$

Experimentally, these strangeness-changing decays are quite weak

$$G_F' \sim \frac{1}{10} G_F \tag{12-7}$$

Consequently, a universality scheme $G_F' = G_F = G_\mu$ is in flat contradiction with experiment. Cabibbo proposed instead that

$$\text{Cabibbo universality: } \sqrt{G_F^2 + (G_F')^2} = G_\mu. \tag{12-8}$$

A convenient way of parametrising this is to put

$$\begin{aligned} G_F &= G_\mu \cos\theta_V \\ & \quad\quad \theta_V = \text{"Cabibbo angle"} \\ G_F' &= G_\mu \sin\theta_V \end{aligned} \tag{12-9}$$

The subscript V stands for vector current, which, as was seen in Chapter 10, is responsible for Fermi transitions. Since the $|\Delta S| = 1$ decays are comparatively feeble we expect θ_V to be small. Comparing the rates for $K^+ \to \pi^0 e^+ \nu_e$ and $\pi^+ \to \pi^0 e^+ \nu_e$ (both pure Fermi $0^- \to 0^-$ transitions) gives a value for $G_G'/G_F = \tan\theta_V$, from which

$$\theta_V = 0.247 \pm 0.008. \tag{12-10}$$

θ_V measures the relative strength of strangeness changing and strangeness conserving decays, as well as accounting for the discrepancy between G_F and G_μ.

A remarkable thing is that comparison of the rates for $K^+ \to \mu^+ \nu_\mu$ and $\pi^+ \to \mu^+ \nu_\mu$, which are *both Gamow–Teller transitions* (see equation (17) below) with coupling constants G_{GT}' and G_{GT} gives

$$\theta_A = \tan^{-1}(G_{GT}'/G_{GT}) = 0.226 \pm 0.005 \tag{12-11}$$

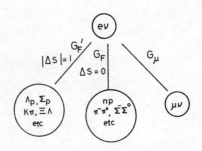

Figure 12.1 Coupling constants for μ decay, and Fermi transitions in $\Delta S = 0$ and $|\Delta S| = 1$ semileptonic decays

which is very close to θ_V. This implies that the Cabibbo angle measures not only the relative strength of Fermi (vector current) $\Delta S = 0$ and $|\Delta S| = 1$ transitions, but also of Gamow--Teller (axial vector current) $\Delta S = 0$ and $|\Delta S| = 1$ transitions:

$$\frac{G'_{GT}}{G_{GT}} \approx \frac{G'_F}{G_F} \quad ; \theta_A \approx \theta_V \quad (12\text{-}12)$$

The baryon decays $\Lambda \to pe^-\nu$, $\Sigma^- \to ne\nu$, $\Sigma^- \to n\mu\nu$ etc, contain both Fermi and Gamow--Teller terms, so may be used to determine θ_V and θ_A separately, giving

$\theta_V = 0.233 \pm 0.012$

$\theta_A = 0.238 \pm 0.018.$ \hfill (12-13)

As a mean value we may take

$\theta_V \approx \theta_A = \theta = 0.25 = 14°.$ \hfill (12-14)

Using (14)

$G_\mu = G_F \sec\theta = 1.458 \times 10^{-49}$ erg cm^3

which is slightly too high, but the errors involved in the estimation of all these numbers are not yet sufficiently small to permit a reliable comparison of theory and experiment.

We should note in passing that Cabibbo's theory can only make sense if G'_F is unaffected by the cloud of virtual particles (kaons in this case). This in turn is guaranteed if the *V-spin current is conserved*; and this is true in the SU$_3$ symmetry limit. In fact in analogy with (10-30) and (10-32) we have

$$\frac{\partial}{\partial x_\lambda} V_\lambda^{K^-\pi^0} \propto m^2(K^-) - m^2(\pi^0)$$

and conservation of $V_\lambda^{K^-\pi^0}$ holds if $m_K = m_\pi$; i.e. in the limit of exact SU$_3$.

12.2 AN INTERPRETATION OF CABIBBO'S THEORY

Equations (8) and (9) may be represented as in Figure 2. The weak interaction vector current of the hadrons has components in the $\Delta S = 0$ and $\Delta S = 1$ "directions" which couple to $(e\nu)$ with strength G_F and G'_F respectively. The coupling strength of the unresolved hadronic current is

$\sqrt{G_F^2 + G_{F'}^{'2}}$

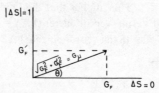

Figure 12.2 The coupling of the hadronic weak current, according to Cabibbo's theory

the same as the coupling strength G_μ of the leptonic current $(\mu\nu)$ in μ decay. Cabibbo's theory is remarkable for the connection it makes between leptons and hadrons — one of the precious few in present day theories.

In the limit of exact SU_3 symmetry, there is no distinction between strange and non-strange particles, and therefore between $\Delta S = 0$ and $|\Delta S| = 1$ processes. The angle θ would therefore be unmeasurable. It is the "medium-strong" symmetry *breaking* interaction which distinguishes isomultiplets of different strangeness, and therefore "fixes" the axes in Figure 2. In this sense, the angle θ is to be blamed more on the strong interaction than on the weak. The origin of Cabibbo's angle is a question of great interest, and as yet an unsolved one. One of the difficulties is that the angle must arise from an interplay between strong and weak interactions, neither of which is really understood.

Another way of visualising the significance of Cabibbo's angle is to consider the leptonic decays of quarks. Assuming that their masses obey $m_s, m_d > m_u$, then the following decays may take place

$$\Delta S = 0, \Delta I = 1: G_F: d \to u + e^- + \bar{\nu}_e$$

$$\Delta S = 1, \Delta I = \tfrac{1}{2}: G_F': s \to u + e^- + \bar{\nu}_e$$

Now turn off the SU_3 breaking interactions, so that d and s become degenerate, but leave electromagnetism turned on, so that they are heavier than u. Then the mixed particle $(\cos\theta d + \sin\theta s)$ decays into u with an amplitude proportional to

$$\cos\theta G_F + \sin\theta G_F' = G_\mu$$

whereas the orthogonal particle $(\sin\theta d - \cos\theta s)$ decays into u with an amplitude proportional to

$$\sin\theta G_F - \cos\theta G_F' = G_\mu(\sin\theta \cos\theta - \cos\theta \sin\theta)$$

$$= 0;$$

in other words, is stable. From the point of view of the weak interactions, these are the natural choice of particles — one decays with an amplitude of "universal" strength, and the other one is stable. It is the medium strong SU_3 breaking interaction which splits up these particles, causing both the resulting "physical" particles (if we may call the quarks physical!) to decay.

These interpretations of the angle θ are due to Cabibbo.

12.3 PARTIALLY CONSERVED AXIAL VECTOR CURRENT

We have seen in Chapter 10 and in section 1 of this chapter that the weak leptonic decays of hadrons are described by a vector current (relativistic form of the Fermi coupling) and an axial vector coupling (relativistic form of the Gamow–Teller coupling). We have also seen that the vector current is conserved, the reason being that the (weak) vector current is identical with the (strong) isospin current. There is then a correspondance between the weak and strong interaction phenomena

$\Delta S = 0$ decays: CVC $\leftrightarrow$ isospin symmetry (SU_2)

$|\Delta S| = 1$ decays: CVC $\leftrightarrow$ unitary spin symmetry (SU_3) (12-15)

It is now natural to ask the question *is the axial vector current also conserved*? And, if so, is there a symmetry manifestation in the strong interactions, analogous to (15)?

The reason that the development of the subject of chiral symmetries did not take place until almost 10 years after the CVC theory of 1958, is that at first sight there are no encouraging answers to the above questions. What physicists now believe, however, is that the axial vector current *is* approximately conserved, and that this has manifestations in strong interactions, but that these are completely different in kind from the manifestations of CVC. A conceptual advance has been required, and it is this that has accounted for the delay.

Let us suppose that the axial vector current is conserved, and call this the CAC hypothesis. Then since the axial current has opposite parity to the vector current, we should expect that to every isospin or SU_3 multiplet in (15) there would exist a *degenerate* multiplet, identical except for having opposite parity. For example, there should be an isodoublet corresponding to the nucleon, with $B = 1, S = 0, I = ½$, mass = 939 MeV and $J^P = ½^-$. This, however, is certainly not true in nature, and not even approximately true. The nucleon has a mass of 939 MeV, but the lowest lying $½^-$ particle with the same quantum numbers has a mass of 1535 MeV. Similarly the pion has a mass of 135 MeV, but the otherwise similar 0^+ particle of lowest mass lies at 975 MeV. This distinct *lack of parity doubling* may be taken as a reason for dismissing the CAC hypothesis. There is another reason for doubting CAC, which was advanced by J. C. Taylor in 1958. This arises from a consideration of the dominant decay of the charged pion

$$\pi^- \to \mu^- + \bar{\nu}_\mu. \tag{12-16}$$

As usual, the interaction is described by a product of currents, one for the $(\mu\nu)$ pair, and one for the pion. The pion current must be a 4-vector, and since π has zero spin, the only 4-vector associated with it is its momentum p_λ. The current is therefore

$$A_\lambda^{\pi^-} = f_\pi p_\lambda \varphi_{\pi^-} \tag{12-17}$$

where f_π is a constant denoting the strength of the interaction (analogous to G_F and G_{GT}) and φ_{π^-} is the pion field. As indicated, the current is an axial vector, since p_λ is a vector, and φ a pseudoscalar. It should be compared with the vector current in (10-30) corresponding to the (rarer) decay $\pi^- \to \pi^0 + e^- + \bar{\nu}_e$. The divergence of $A_\lambda^{\pi^-}$ is calculated in the same way that (10-32) was derived; the answer is

$$\sum_\lambda \frac{\partial}{\partial x_\lambda} A_\lambda^{\pi^-} = i m_\pi^2 f_\pi \varphi_{\pi^-}. \tag{12-18}$$

If CAC holds, this should be zero. Since $m_\pi \neq 0$, this implies that $f_\pi = 0$; in this case *pion decay is forbidden*. This, of course, is nonsense, and the axial vector current is therefore not conserved.

Faced with these two rather daunting facts (no parity doubling, pion decay not forbidden), this would seem to be the end of the matter. A number of people, however, notably Nambu and Adler, did not let the question rest, and argued along the following lines.

The right hand side of (18) is not zero, but m_π^2 is *small* — the pion is the lightest hadron there is — so the divergence is *almost* zero. Equation (18) is taken to be central in defining the hypothesis of *partial conservation of the axial vector current* (PCAC): the axial current is not conserved, but it would be if the pion had zero mass; it is "partially" conserved. The question now is, what consequences does this have? How are the two difficulties above avoided? We have seen that parity doubling is not even approximately ("partially") true, so we must look for an alternative consequence. The clue is to compare the currents $V_\lambda^{\pi^-\pi^0}$ and $A_\lambda^{\pi^-}$ in $\pi^- \to \pi^0 e^- \bar{\nu}_e$ and $\pi^- \to e^- \bar{\nu}_e$ decays respectively. It is easiest to set out the comparison in a table — see Table 1. The first few entries should be clear, and have for the most part been mentioned above. The currents V_λ and A_λ induce transitions between the *hadron* states $\pi^- \to \pi^0$ and $\pi^- \to$ vacuum, and they are conserved if these have equal mass; CVC implies $m(\pi^-) = m(\pi^0)$ and PCAC implies that *if* $m_\pi = 0$ then the axial current would be conserved. Now we have seen above that CVC is related to isospin symmetry. For example, the condition $m(\pi^-) = m(\pi^0)$ holds if isospin symmetry is exact. As a further example, isospin relates the amplitudes for processes involving π^- to those involving π^0 by simple Clebsch–Gordon coefficients; for instance it follows from equation (3A-27) that

$$\text{Amp}\,(\pi^- p \to \pi^0 n) = \sqrt{2}\,\text{Amp}\,(\pi^- p \to \pi^- p)$$

near the $\Delta(1236)$ resonance (i.e. if scattering takes place in an $I = \tfrac{3}{2}$ state). Similarly, for the axial current A_λ, in the limit in which it is conserved there is a relation between processes involving (massless) pions and those *not* involving pions. So in the realistic case of *partial* conservation, there is a relation between processes involving (massive) pions and those not involving pions. That is, there is a relation between the amplitude for the process $a \to \beta$ (a and β are arbitrary) and the process $a \to \beta + \pi^0$; and thence $\pi + a \to \pi + \beta$, $a \to \beta + 2\pi$ etc. Moreover, these relations become exact in the limit in which π and "—", the vacuum, are degenerate, i.e. in which the pion has *zero energy and momentum* (and therefore, of course, zero mass). Such pions are called *soft*. In other words, conservation of the axial vector current would result, not in degeneracies, but in *relations between the amplitudes for processes involving different numbers of soft pions*. These relations are examples of the so-called *Adler consistency conditions*.

In the real world pions are not massless, so the Adler condition has to be accompanied by a prescription for taking into account the mass of the pion. The prescription used is that as the pion mass is changed from its physical value to zero, the scattering amplitude varies *smoothly*. Mathematically this is expressed by extrapolation using dispersion relation methods. For example, the Adler condition for π-N scattering says that the cross section for soft pions should be zero

$$\sigma(\pi + N \to \pi + N) \xrightarrow{p_\lambda^\pi \to 0} 0. \qquad (12\text{-}19)$$

Table 12.1

Current	Transition	Current conserved if	Current conservation implies a relation between processes involving the strong interactions of
$V_\lambda^{\pi^-\pi^0}$	$\pi^- \to \pi^0 (+ e\nu)$	$m_{\pi^-} = m_{\pi^0}$	π^- and π^0
$A_\lambda^{\pi^-}$	$\pi^- \to - (\mu\nu)$	$m_{\pi^-} = 0$	π^- and — (vacuum)

This is a remarkable prediction. Using the experimental data, all that may be deduced is the value of

$$\lim \sigma(\pi + N \to \pi + N_j). \tag{12-20}$$

where the limit taken is $\mathbf{p}^\pi \to 0, p_4 \to im_\pi$.

Adler showed, using dispersion relations, that the limit of (20) as $m_\pi \to 0$ is in fact consistent with zero.

Note that this condition involves only strong interactions, and it is a symmetry condition corresponding to the partial conservation of the weak axial vector current. Nambu has termed this symmetry a *hidden symmetry*, since even when the symmetry is exact, physical states are *not eigenstates* of the symmetry operators. For example, by virtue of isospin symmetry, particles are eigenstates of

$$I_3 p = \tfrac{1}{2} p$$

but the corresponding operator derived from the axial current, which is denoted I_3^5, has the effect

$$I_3^5 p = (p + \pi^0)$$

so neither the proton nor any other particle is in an eigenstate of I_3^5 — just as no particle is in an eigenstate of I_+ or I_-. (The superscript 5 comes from γ_5, the matrix involved in the axial vector current of a Dirac particle — see equation (10-7).)

12.4 CHIRAL SYMMETRY

Isospin transformations generate the rotation group SU_2. How is this group enlarged when we take PCAC into account? Just as the isospin charges are defined by

$$I_i = \int V_0^i d^3 x \ (i = 1, 2, 3) \tag{12-21}$$

where $V_0^i = -iV_4^i$ is the time component of the $\Delta I = 1$ vector current, so the "axial isospin" charges are

$$I_i^5 = \int A_0^i d^3 x \ (i = 1, 2, 3) \tag{12-22}$$

where A_0^i is the time component of the $\Delta I = 1$ axial current. I_1, I_2 and I_3 are the generators of isospin rotations, so they obey the commutation relations (2-49).

CABIBBO'S THEORY

$$[I_1, I_2] = iI_3 \text{ \& cyclic} \tag{12-23}$$

I_i^5 are the components of a vector in isospin space, so it follows that the commutation relations of I_i^5 with I_i are

$$[I_1^5, I_2] = iI_3^5 \text{ \& cyclic.} \tag{12-24}$$

This has a parallel with the commutation relations between momentum and angular momentum (2-50), expressing the fact that momentum is a vector in ordinary space.

To complete the picture, we need to know how the components of I^5 commute amongst themselves. One possibility is that

$$[I_1^5, I_2^5] = 0 \text{ etc} \tag{12-25}$$

and this is suggested by various field theory models of the axial vector current. On the basis of the quark model, however, Gell-Mann suggested that

$$[I_1^5, I_2^5] = iI_3 \text{ \& cyclic.} \tag{12-26}$$

The commutation relations (23), (24) and (26) may be re-expressed by defining

$$I_i^L = \tfrac{1}{2}(I_i + I_i^5), \quad I_i^R = \tfrac{1}{2}(I_i - I_i^5) \tag{12-27}$$

called "left-handed" and "right-handed" isospins. It is then straightforward to show that

$$[I_1^L, I_2^L] = iI_3^L \text{ \& cyclic}$$

$$[I_1^R, I_2^R] = iI_3^R \text{ \& cyclic}$$

$$[I_i^L, I_j^R] = 0. \tag{12-28}$$

I_i^L and I_i^R are so called because in the quark model they act on the left- and right-handed quark wave functions. Because I and I^5 have opposite parity, I^L and I^R are transformed into each other by parity

$$I^L \overset{P}{\leftrightarrow} I^R \tag{12-29}$$

I^L and I^R generate two rotation groups that commute with each other. The full group is then the product of groups $SU_2 \times SU_2$.

Strong interactions are approximately invariant under $SU_2 \times SU_2$. To be more precise, in the limits in which electromagnetism and the pion mass are both negligible, strong interactions are $SU_2 \times SU_2$ invariant, and we then have

$$\frac{d}{dt} I^L = \frac{d}{dt} I^R = 0$$

or, equivalently

$$m_\pi = 0 \quad [H, I^L] = [H, I^R] = 0 \tag{12-30}$$

This symmetry is an example of *chiral symmetry*. (Chirality means handedness).

The whole scheme may be extended to SU_3. Let the generators of SU_3 be labelled F_i, which are a generalisation of I_1, I_2 and I_3. There are 8 of them (I-spin, U-spin and V-spin make 9, with 1 relation between them; $9 - 1 = 8$), so $i = 1, \ldots, 8$. SU_3 is exact when the SU_3 multiplets are degenerate — when the so-called medium strong interactions are turned off. PCAC may hopefully be generalised to SU_3, and then in parallel with (18) we would have for the $|\Delta S| = 1$ axial current

$$\sum_\lambda \frac{\partial}{\partial x_\lambda} A_\lambda^K = f_K m_K^2 \varphi_K$$

where f_K is the constant measuring the strength of $K^- \to \mu^- \bar{\nu}_\mu$, m_K is the kaon mass and φ_K the kaon wave function. This generalised PCAC would then result in Adler conditions for soft kaons, applicable properly when $m_K = 0$. To verify these conditions, therefore, requires an extrapolation from m_K to 0, over almost 500 MeV, and this is difficult to do unambiguously.

Defining the axial charges of SU_3 by F_i^5, we again take the combinations

$$F_i^L = \tfrac{1}{2}(F_i + F_i^5), \quad F_i^R = \tfrac{1}{2}(F_i - F_i^5)$$

generating left-handed and right-handed unitary spins. They define together the group $SU_3 \times SU_3$. This is an exact only when

1) electromagnetism is absent,
2) SU_3 is exact, i.e. the medium strong breaking interaction is absent,
3) $m_\pi = m_K = m_\eta = 0$.

The real world bears very little relation to this! Consequently $SU_3 \times SU_3$ is a badly broken symmetry. These results are summarised in Table 2.

Table 12.2

Weak current	Conservation hypothesis	Exact in limit	Strong interaction symmetry			Exactness of symmetry
			Name	Group	Manifestation	
$V \Delta S=0$	CVC	no el. mag.	isospin symmetry	SU_2	degenerate isospin multiplets	almost exact
$A \Delta S=0$	PCAC	$m_\pi = 0$	chiral isospin	$SU_2 \times SU_2$	degenerate isospin multiplets zero mass pions Adler cond. for soft π	strongly broken
$V \Delta S=0$ & $V \Delta S=1$	CVC	no el. mag. no medium strong intn.	unitary symmetry	SU_3	degenerate SU_3 multiplets	strongly broken
$A \Delta S=0$ & $A \Delta S=0$	PCAC	$m_\pi = m_K = m_\eta = 0$	chiral unitary symmetry	$SU_3 \times SU_3$	degenerate SU_3 multiplets zero mass π, K, η Adler cond. for soft π, K, η	very strongly broken

12.5 CURRENT ALGEBRA

Finally we turn to the question of the value of $g_A = G_{GT}/G_F$, which was calculated in 1965 by Adler and Weisberger using the so-called current algebra proposed by Gell-Mann, and the PCAC theory, described above. The Adler–Weisberger sum rule, as it is called, is remarkable for two reasons; (1) the predicted value of g_A is very close to the experimental value, (2) g_A turns out to be related to, of all things, the pion-nucleon scattering cross section. This is another example of a connection between weak and strong interactions. Only the barest outline can be given below of the calculation of Adler and Weisberger, which in its details lies far beyond the scope of this book.

First, however, let us examine the assumptions of current algebra, and for simplicity take the case of $SU_2 \times SU_2$. We saw above that if nature is $SU_2 \times SU_2$ symmetric the following relations hold. First, the $SU_2 \times SU_2$ transformations are generated by $I_1, I_2, I_3, I_1^5, I_2^5, I_3^5$, which have the commutation relations (23), (24) and (26)

$$[I_1, I_2] = iI_3 \ \& \text{ cyclic}$$

$$[I_1^5, I_2] = iI_3^5 \ \& \text{ cyclic}$$

$$[I_1^5, I_2^5] = iI_3 \ \& \text{ cyclic}. \tag{12-31}$$

Second, since these generators give rise to a (broken) symmetry, the generators are (approximately) conserved, that is

$$[I_i, H] = [I_i^5, H] \approx 0 \ (i = 1, 2, 3). \tag{12-32}$$

Now the outstanding problem connected with isospin, unitary spin and chiral symmetries is precisely that they are *not exact symmetries;* that is, that (32) is only approximate. Gell-Mann suggested, therefore, as early as 1962, that *relations (31) hold exactly, even though the relations (32) are only approximate.* That is, we may get as much information as possible from (31), and not worry that the symmetry is only approximate; the results obtained should be independent of this. The attraction of this scheme is that, if successful, it gives a mathematically precise meaning to an approximate symmetry. In this sense the philosophy of current algebra is to take a step beyond symmetry, but using concepts developed in connection with symmetry. The scheme is called current algebra because the relations (31), which hold for the generators of a group, are called the *algebra* of the generators; and in this case the generators are related to physical currents.

The Adler-Weisberger relation was the first outstanding success of current algebra. It is derived from the third set of equations (31), which, it will be recalled, was a separate postulate (see (26) above), which is not necessarily true (see (25)), but which is suggested by the quark model. If it holds, then the algebra (or group) is that of $SU_2 \times SU_2$ (see (27) and (28)). Defining as usual

$$I_\pm^5 = I_1^5 \pm i I_2^5 \tag{12-33}$$

then the third set of equations (31) gives

$$[I_+^5, I_-^5] = 2I_3. \tag{12-34}$$

The matrix elements of both sides of (34) are now evaluated between physical proton states. This gives a "sum rule"

$$\sum_a (\psi_p^* I_+^5 \psi_a)(\psi_a^* I_-^5 \psi_p) - \sum_a (\psi_p^* I_-^5 \psi_a)(\psi_a^* I_+^5 \psi_p) = 2(\psi_p^* I_3 \psi_p). \tag{12-35}$$

There is a summation over all intermediate states a. Since $(\psi_p^* I_+^5 \psi_a) = (\psi_a^* I_-^5 \psi_p)^*$, (35) becomes

$$\sum_a |\psi_a^* I_-^5 \psi_p|^2 - \sum_a |\psi_a^* I_+^5 \psi_p|^2 = 2\psi_p^* I_3 \psi_p. \tag{12-36}$$

The right hand side of (36) is clearly unity. In the first term on the left, one of the important intermediate states is the neutron, since I^5 lowers the charge of p, and the weak current responsible for the $p \to n$ transition contains an axial, as well as a vector term. Its coefficient is proportional to G_{GT}. (36) then becomes (with another assumption)

$$\left(\frac{G_{GT}}{G_F}\right)^2 + \sum_{a \neq n} |\psi_a^* I_-^5 \psi_p|^2 - \sum_a |\psi_a^* I_+^5 \psi_p|^2 = 1. \tag{12-37}$$

At this point PCAC is used. From (22) $I_+^5 = \int A_0^+ d^3x$, and from (18)

$$\sum_\lambda \frac{\partial}{\partial x_\lambda} A_\lambda^+ = \text{const. } \varphi_{\pi+};$$

the overall effect is that $I_\pm^5$ is related to the $\pi^\pm$ fields. The last term on the right is then proportional to the $\pi^+ p$ scattering cross section, and the previous term to the $\pi^- p$ scattering cross section. The sum rule is now

$$1 - \frac{1}{g_A^2} = \text{const.} \int \frac{dW}{W} [\sigma^{\pi^+ p}(W) - \sigma^{\pi^- p}(W)], \tag{12-38}$$

where W is the pion energy. An interesting property of (38) is that its right hand side is the difference between two positive quantities, which could in principle be positive or negative, so there is no indication whether $G_{GT} > G_F$ or $G_{GT} < G_F$. In fact, most of the contribution to the integral comes from the $\Delta(1236)$ resonance, and as one can see at a glance from Figure 8-1, in the region of this resonance

$$\sigma^{\pi^+ p} \gg \sigma^{\pi^- p}$$

so the rhs is positive, and

$$|g_A| > 1.$$

Since PCAC was formulated for zero mass pions, a correction has to be made for the finite pion mass, and Weisberger arrived at the prediction

$$|g_A| = 1.15$$

to be compared with the experimental value

$$g_A = 1.18 \pm 0.02.$$

The agreement is excellent, and the problem of the curious value of g_A is solved by relating it to the *dynamics of pion-nucleon scattering*. In parenthesis, it is also worth noting that the success of the Alder–Weisberger sum rule constitutes that the commutator of two axial charges is a vector charge, i.e. that equation (14) rather than (13) holds true, and therefore that the current algebra is that of $SU_2 \times SU_2$. This may be thought of as indirect evidence for the quark model.

The scheme may be extended to $SU_3 \times SU_3$, by relating g_A to the KN and $\overline{KN}$ scattering cross sections. There are, as mentioned above, more uncertainties involved in this calculation, but Weisberger found that

$$\dot{g}_A = 1.20 \pm 0.10$$

which is certainly consistent with the previous value.

FURTHER READING

Adler, S. L. and R. F. Dashen *Current Algebras,* Benjamin, 1968.

Cabibbo, N. "Unitary Symmetry and Leptonic Decays", *Physical Review Letters,* **10**, 531 (1963).

Cabibbo, N. "Weak Interactions and the Unitary Symmetry", in *Weak Interactions and High Energy Neutrino Physics,* "Enrico Fermi" International School of Physics, course 32, Academic Press, 1966.

Gell-Mann, M. "The Symmetry Group of Vector and Axial Vector Currents", *Physics,* **63** (1964). Reprinted in M. Gell-Mann and Y. Ne'eman, *The Eightfold Way,* Benjamin, 1964.

Gell-Mann, M., R. J. Oakes and B. Renner, "Behavior of Current Divergences under $SU_3 \times SU_3$", *Physical Review,* **175**, 2195 (1968).

Nambu, Y. "Chiral Symmetries", in *Group Theoretical Concepts and Methods in Elementary Particle Physics* (F. Gürsey, ed.), Gordon and Breach, 1964.

Renner, B., *Current Algebras and their Application,* Pergamon, 1968.

Chapter 13

Unified Weak and Electromagnetic Interactions and Charm

It has been realised for many years that there is one respect in which the weak and electromagnetic interactions are alike, namely, that if the intermediate boson W exists, then, like the photon, it will have spin 1. There is, however, another respect in which they differ, and that is that whereas quantum electrodynamics is "renormalisable", Fermi's theory of weak interactions is "non-renormalisable". What this means in crude terms is that a perturbation series in the Fermi interaction turns out to be hopelessly divergent, so that each term is infinitely greater than the previous one! This is not the case for electrodynamics and the reason for this depends on the fact that the photon has spin 1 and is massless. (That it is massless is intimately related to the fact that electrodynamics is invariant under gauge transformations.) In the last few years a "unified" theory of weak and electromagnetic interactions has been developed, principally by Weinberg and Salam, which exploits their similarity (spin 1 for the quanta of both fields) to provide a *renormalisable* model of weak interactions.

The difficulty which has had to be surmounted is that although the intermediate boson has spin 1, it is not massless. The theory has the considerable aesthetic attraction that in it the weak and electromagnetic interactions are not independent — they are merely "components" of *one* interaction.

There are many variants of the original Weinberg-Salam Theory, but a characteristic feature of most of them is the prediction of so-called weak *neutral currents*. Evidence for some of these currents has indeed been found. At the same time, however, other predicted neutral currents seem not to exist. A way round this difficulty has been suggested by invoking the existence of 4 quarks instead of 3. The fourth quark would carry a new quantum number, called "charm".

This chapter is devoted to a discussion of these matters, and ends with a note on the new particles discovered in November 1974, which are at present (April 1975) not understood, but which some physicists believe constitute evidence for charm.

13.1 UNIFIED THEORY OF WEAK AND ELECTROMAGNETIC INTERACTIONS

As mentioned above, the motive for a unified theory is to arrive at a theory of weak interactions which is "renormalisable", i.e. which gives (after redefinition of some constants) finite cross sections in perturbation theory. In this section I shall discuss the conditions which the weak interaction must fulfill for this to be the case.

The best way to start is to turn back to section 1.3, in particular figures 1.4, 1.6 and 1.7. Figure 1.4 represents the electromagnetic interaction between e^- and p, 1.6 represents Fermi's "point" interaction responsible for neutron beta-decay, and 1.7 represents the same process mediated by an intermediate boson W^-. In classical field theory the electromagnetic interaction is described by saying that a charged particle creates about itself an electromagnetic *field*, and the motion of any other charged particle in this field will be duly affected. In quantum theory, the field has to quantised, i.e. to have a particle aspect, and the quantum of the electromagnetic field is the photon γ – this is figure 1.4. If the weak interaction works the same way, then particles with weak interactions will create about themselves a "weak field", and other particles with weak interactions will respond to this field. The quantum of this weak field is W – this is figure 1.7 (the argument of course parallels also Yukawa's argument that there is a strong field, with quantum π, as in figure 1.5).

We saw in section 10.2 that in Fermi's theory the coupling constant of the $n-p-e-\nu$ vertex of figure 1.6 is denoted G_F. If, however, beta-decay really takes place according to figure 1.7, then the relevant coupling constant is that of the $p-n-W$ (or $e-\nu-W$) vertex, denoted g. In an analgous way, the coupling constant of the $p-p-\gamma$ and $e-e-\gamma$ vertices of figure 1.4 is simply electric charge e. g may be said to be the amount of "weak charge" carried by the decaying particles.

$$\left.\begin{array}{l} n-p-e-\nu \text{ coupling constant } G_F \\[4pt] n-p-W \quad \text{coupling constant } g \\[4pt] p-p-\gamma \quad \text{coupling constant } e \end{array}\right\} \tag{13.1}$$

Since figures (1.6) and (1.7) describe, overall, the same process, then there must be a relation between G_F and g. This relation is

$$G_F = \frac{g^2}{4\sqrt{2} M_W^2} \tag{13.2}$$

The smallness of G_F may then be attributed either to a small g ("intrinsically weak interaction") or to a large mass M_W. If M_W is large enough g may be quite large, and the weak interactions would then appear weak only because the intermediate boson is so heavy. Equation (13.2) may be rewritten to give

$$M_W^2 = \frac{g^2}{4\sqrt{2}G_F} \tag{13.3}$$

A *unified* theory of weak and electromagnetic interactions implies a relation between e and g, which, through (3), then gives a prediction for M_W. We shall return to this point later.

Let us now collect together what we have learned in previous chapters about semi-leptonic decays of hadrons — all hadrons, not just the neutrons. First of all, we saw in chapter 5 (equations (5.16) and (5.32), pages 89 and 92) that there are two sorts of decay, one with $\Delta S = 0$, $\Delta I = 1$ and the other with $\Delta S = 1$, $\Delta I = \frac{1}{2}$. Further, we saw in chapter 12 (equation (12.9) page 210) that the relevant Fermi coupling constants for these decays were $G\cos\theta$ and $G\sin\theta$ respectively, where $G = G_\mu$ and $\theta = \theta_V = \theta_A$ = Cabibbo angle. Next, we saw in chapter 10 (equations (10.8), (10.10) page 173) that the Fermi interaction is written in a "current–current" form, so collecting these results together we may generalise (10.10) to write for the generic Fermi interaction Hamiltonian

$$H = G\left[\underbrace{\cos\theta J_+^{\Delta S=0} + \sin\theta J_+^{\Delta S=1}}_{\text{hadron currents}}\right]\left[\underbrace{j_-}_{\text{lepton current}}\right] \tag{13.4}$$

where the lepton current is given by

$$j_- = \bar{e}\nu \tag{13.5}$$

(13.5) is shorthand for the second factor in (10.10); I have written $\bar{e}$ instead of $\bar{\psi}_e$, ν instead of ψ_ν and omitted the factor $\gamma_\lambda(1 + \gamma_5)$. The minus sign in (13.5) indicates that the current is negative; it creates an electron e^- and annihilates a neutrino ν (or, what is equivalent, creates an antineutrino $\bar{\nu}$). Similarly, the hadron current in (13.4) is positive. For example, neutron decay creates a proton and annihilates a neutron, π^- decay (see equation (12.5)) annihilates π^- and creates π°. In both cases the hadronic charge increases by 1 unit. Similarly typical components of $J_+^{\Delta S=1}$, like $\Lambda \to p + e^- + \bar{\nu}$, $\Sigma^- \to n + e^- + \bar{\nu}$ involve an increase in hadronic charge.

Finally, we express equation (4) in terms of the quark model, for which I refer to chapter 11, and in particular to Figure 11.2 (page 193). Consider $J_+^{\Delta S=0}$. It will contain terms like $n \to p$, $\pi^- \to \pi^\circ$, $\Sigma^\circ \to \Sigma^+$ etc. Writing out the quark content of these particles we have $n = (udd)$, $p = (uud)$, $\pi^- = (\bar{u}d)$, $\pi^\circ = 1/\sqrt{2}\,(\bar{u}u - \bar{d}d)$ etc, as may be checked, for example, from table 11.1 and figure 11.3. The result is that *all* terms in $J_+^{\Delta S=0}$ are equivalent to the quark transition $d \to u$ (or $\bar{u} \to \bar{d}$). Thus we may write

$$J_+^{\Delta S=0} = (\bar{u}d) \tag{13.6}$$

which destroys a d quark and creates a u quark (or, equivalently, destroys a $\bar{u}$ and creates a $\bar{d}$). In a similar way, for the strangeness changing current we have, in quark language

$$J_+^{\Delta S=1} = (\bar{u}s) \tag{13.7}$$

in which a u quark is created from an s quark. Combining now equations (4)–(7) we have

$$H = G[\cos\theta\,\bar{u}d + \sin\theta\,\bar{u}s]\,[\bar{e}\nu] \tag{13.8}$$

$$\equiv G[J_+]\,[j_-]$$

The reader may by this stage have forgotten that we wish to write down the conditions that the weak interaction is renormalisable, like the electromagnetic one. We are now in a position to do this. It was conjectured by Steven Weinberg in 1967 (and later proved by G. 't Hooft) that the weak interaction is renormalisable if

(a) the "bare" mass of W is zero, just like γ
(b) the hadron and lepton currents J and j are the generators of a symmetry group.

Condition (a) is rather technical. It amounts to saying that in an "ideal" world, $M_W = 0$; the intermediate boson then acquires a non-vanishing mass by a rather freakish trick in quantum field theory, which goes by the name of the "Higgs mechanism". This will be described further in section 3, but will not concern us for the present.

It is condition (b) which we shall now concentrate on. In principle the group could be any group, but in the Weinberg-Salam model it is chosen to be the three-dimensional rotation group*. We saw in chapter 2 that the generators of rotations in ordinary three-dimensional space are (equation (2.17)) L_x, L_y and

---

* To be precise, the group is the product group SU(2) × U(1). SU(2) is the rotation group generated by "weak isospin" and U(1) is an abelian gauge group generated by "weak hypercharge".

L_z which obey the relations (2.49) (page 39). It follows that j must have three components (though this time in an abstract space) which must also obey (2.49):

$$j_x j_y - j_y j_x = ij_z$$

$$j_y j_z - j_z j_y = ij_x$$

$$j_z j_x - j_x j_z = ij_y \qquad (13.9)$$

Putting

$$j_+ = j_x + ij_y, \, j_- = j_x - ij_y, \, j_0 = j_z,$$

this gives

$$j_+ j_- - j_- j_+ \equiv [j_+, j_-] = 2j_0 \qquad (13.10)$$

Let us now calculate j_0. From (8) we have $j_- = \overline{(ev)}$, and we express this in matrix form as follows

$$j_- = \overline{ev}$$

$$= (\overline{v}\,\overline{e}) \begin{pmatrix} 0 & 0 \\ 1 & 0 \end{pmatrix} \begin{pmatrix} v \\ e \end{pmatrix} \qquad (13.11)$$

$$\equiv \overline{\psi} m_- \psi$$

where

$$\psi = \begin{pmatrix} v \\ e \end{pmatrix}, \, \overline{\psi} = (\overline{v}\,\overline{e}), \, m_- = \begin{pmatrix} 0 & 0 \\ 1 & 0 \end{pmatrix}.$$

Similarly

$$j_+ = \overline{v}e$$

$$= (\overline{v}\,\overline{e}) \begin{pmatrix} 0 & 1 \\ 0 & 0 \end{pmatrix} \begin{pmatrix} v \\ e \end{pmatrix}$$

$$\equiv \overline{\psi} m_+ \psi \qquad (13.12)$$

where

$$m_+ = \begin{pmatrix} 0 & 1 \\ 0 & 0 \end{pmatrix}$$

Equations (11) and (12) rely only on the usual rules for matrix multiplication, as the reader will verify. We now have

$$[m_+, m_-] = \begin{pmatrix} 0 & 1 \\ 0 & 0 \end{pmatrix}\begin{pmatrix} 0 & 0 \\ 1 & 0 \end{pmatrix} - \begin{pmatrix} 0 & 0 \\ 1 & 0 \end{pmatrix}\begin{pmatrix} 0 & 1 \\ 0 & 0 \end{pmatrix}$$

$$= \begin{pmatrix} 1 & 0 \\ 0 & -1 \end{pmatrix}$$

so that, from (10), (11) and (12)

$$2j_0 = \bar\psi\,[m_+, m_-]\,\psi$$

$$= (\bar\nu\,\bar e)\begin{pmatrix} 1 & 0 \\ 0 & -1 \end{pmatrix}\begin{pmatrix} \nu \\ e \end{pmatrix}$$

$$= (\bar\nu\nu - \bar e e)$$

$$j_0 = \tfrac{1}{2}(\underset{\text{Weak neutral current}}{\bar\nu\nu} - \underset{\text{Electromagnetic current}}{\bar e e}) \tag{13.13}$$

The condition (b) that the physical currents generate a rotation group implies that if j_+ and j_- are physical currents (and they are) then j_0 must be too. j_0 consists of two terms, the second of which, $\bar e e$, is a current which annihilates one electron and creates another. This is clearly just the electromagnetic current, and corresponds to the e–e–γ vertex in Figure 1.4. To be precise, it could also correspond to the graph of Figure 1.4, but mediated by a (heavy) neutral intermediate boson Z°, instead of a photon. We see, then, that the electromagnetic current is inevitably brought into the symmetry scheme — this is why the theory is a *unified* theory of weak and electromagnetic interactions. However, there is another component to j_0, namely $\bar\nu\nu$, which creates one neutrino and annihilates another, for example as in the processes

$$\nu + p \to \nu + n + \pi^+$$

$$\nu + p \to \nu + p + \pi^\circ \tag{13.14}$$

The theory then predicts that besides the conventional weak processes, like $n \to p + e^- + \bar{\nu}$ or $\nu + n \to p + e^-$, in which one neutrino and one electron appear, there should also exist processes like (14) which involve *neutrinos only*. These are known as *weak neutral current processes*.

After many years of looking for processes like (14), they were at last observed in 1973 at CERN, Geneva, and in 1974 at the Fermi National Accelerator Lab., in the U.S.A. For a discussion of the experiments and references to the original literature I refer the reader to the two review articles by Cline, Mann and Rubbia mentioned at the end.

While chalking up the observation of weak neutral currents as a success for the Weinberg-Salam theory, however, I should bring the reader's attention to the fact that in (14) I indicated, with no justification, that the *hadronic* current, as well as being neutral, also carried *no strangeness*; the hadrons on the right hand side have $S = 0$. The reason for stipulating this stems from the observation that the *neutral current decays* of the K mesons

$$K^+ \to \pi^+ + e^+ + e^-$$

$$K^+ \to \pi^+ + \nu + \bar{\nu}$$

$$K^\circ \to \mu^+ + \mu^- \tag{13.15}$$

do not take place, from which we deduce that *strangeness changing neutral currents do not exist*. With this crucial restriction in mind, we must now calculate the general form for the hadronic neutral current using the fact that the currents generate a rotation group. What we shall find is that both strangeness conserving and strangeness changing currents exist. In order to banish the strangeness changing current, we have to introduce a fourth quark, with a new quantum number, "charm". This is done in the next section. Before leaving this section, however, let me remark that the unification of weak and electromagnetic interactions implied for example by (13) leads to a relation between e, the electric charge, and g the $p-n-W$ coupling constant in (1). This relation is simply that g is greater than e

$$g > e$$

From (3) this gives

$$M_w^2 > \frac{e^2}{4\sqrt{2}G_F} \tag{13.16}$$

It should be understood that (16) holds only in the "natural" system of units when $\hbar = c = 1$. The dimensionally correct equation (remembering that G has units of Jm^3) is

$$M_w^2 > \frac{1}{4\sqrt{2}} \cdot \frac{e^2}{\epsilon_0 \hbar c} \cdot \frac{\hbar^3}{cG} \cdot$$

Here the factor $e^2/\epsilon_0 \hbar c = 4\pi\alpha$ where α is the fine structure constant = 1/137, is dimensionless. Inserting $G = 1.4 \times 10^{-62}\ Jm^3$ gives

$$M_w > 6.68 \times 10^{-26}\ Kg = 37\ GeV \tag{13.17}$$

Hence the Weinberg-Salam theory implies a lower limit on the intermediate boson mass of 37 GeV; this is perfectly consistent with the experimental lower limit of 2 GeV (equation (1.5))!

13.2 CHARM

We calculate J_0, the neutral hadronic current, in the same way that we calculated j_0, the leptonic neutral current. From (8), we have for the positively charged current

$$J_+ = \bar{u} \cos\theta\, d + \bar{u} \sin\theta\, s$$

$$= (\bar{u}\ \bar{d}\ \bar{s}) \begin{pmatrix} 0 & \cos\theta & \sin\theta \\ 0 & 0 & 0 \\ 0 & 0 & 0 \end{pmatrix} \begin{pmatrix} u \\ d \\ s \end{pmatrix}$$

$$\equiv \bar{\psi} M_+ \psi \tag{13.18}$$

where

$$\bar{\psi} = (\bar{u}\ \bar{d}\ \bar{s}),\ \psi = \begin{pmatrix} u \\ d \\ s \end{pmatrix},\ M_+ = \begin{pmatrix} 0 & \cos\theta & \sin\theta \\ 0 & 0 & 0 \\ 0 & 0 & 0 \end{pmatrix}$$

ELEMENTARY PARTICLES AND SYMMETRIES

Similarly the negatively charged current is

$$J_- = \bar{d}\cos\theta\, u + \bar{s}\sin\theta\, u$$

$$= (\bar{u}\ \bar{d}\ \bar{s}) \begin{pmatrix} 0 & 0 & 0 \\ \cos\theta & 0 & 0 \\ \sin\theta & 0 & 0 \end{pmatrix} \begin{pmatrix} u \\ d \\ s \end{pmatrix}$$

$$\equiv \bar{\psi} M_- \psi$$

where

$$M_- = \begin{pmatrix} 0 & 0 & 0 \\ \cos\theta & 0 & 0 \\ \sin\theta & 0 & 0 \end{pmatrix}$$

In analogy with (10), we have then

$$2J_0 = \bar{\psi}[M_+, M_-]\psi$$

$$= (\bar{u}\ \bar{d}\ \bar{s}) \begin{pmatrix} 1 & 0 & 0 \\ 0 & -\cos^2\theta & -\cos\theta\sin\theta \\ 0 & -\cos\theta\sin\theta & -\sin^2\theta \end{pmatrix} \begin{pmatrix} u \\ d \\ s \end{pmatrix}$$

$$= \underbrace{(\bar{u}u - \cos^2\theta\, \bar{d}d - \sin^2\theta\, \bar{s}s)}_{\Delta S = 0} - \underbrace{\sin\theta\cos\theta\, (\bar{d}s + \bar{s}d)}_{|\Delta S| = 1}$$

The hadronic neutral current then contains, besides $\Delta S = 0$ terms, which convert a u quark into a u quark, or d into d, or s into s, also $|\Delta S| = 1$ terms, which convert d into s or vice versa. These are precisely the terms we do not want, because of the non-occurrence of the decays (15). The unified gauge theory taken with the Cabibbo form for the hadron current has run into trouble. A way to avoid this, however, was proposed by Glashow, Iliopoulos and Maiani in 1970 (though their paper was not concerned with the Weinberg-Salam theory). These authors suggested that there are 4 quarks, not 3. The extra quark, c, has the same baryon number, charge and strangeness as u, but differs from it by a hitherto unnoticed quantum number, charm C. The four quarks are displayed in table 13.1. The charged hadron current J_+ now contains, in addition to the $\bar{u}d$ and $\bar{u}s$ terms in (18) ["charm conserving" terms], $\bar{c}d$ and $\bar{c}s$ ["charm changing"] terms:

UNIFIED INTERACTIONS AND CHARM

Table 13–1 Quantum numbers of the 4 quarks in the "charmed quark" model.

Quark	B	Q	S	C
c	$\frac{1}{3}$	$\frac{2}{3}$	0	1
u	$\frac{1}{3}$	$\frac{2}{3}$	0	0
d	$\frac{1}{3}$	$-\frac{1}{3}$	0	0
s	$\frac{1}{3}$	$-\frac{1}{3}$	-1	0

$$J_+ = \bar{u}(\cos\theta\, d + \sin\theta\, s) + \bar{c}(-\sin\theta\, d + \cos\theta\, s)$$

$$= (\bar{c}\ \bar{u}\ \bar{d}\ \bar{s}) \begin{pmatrix} 0 & 0 & -\sin\theta & \cos\theta \\ 0 & 0 & \cos\theta & \sin\theta \\ 0 & 0 & 0 & 0 \\ 0 & 0 & 0 & 0 \end{pmatrix} \begin{pmatrix} c \\ u \\ d \\ s \end{pmatrix}$$

$$\equiv \bar{\psi} C_+ \psi \tag{13.19}$$

Similarly

$$J_- = (\bar{c}\ \bar{u}\ \bar{d}\ \bar{s}) \begin{pmatrix} 0 & 0 & 0 & 0 \\ 0 & 0 & 0 & 0 \\ -\sin\theta & \cos\theta & 0 & 0 \\ \cos\theta & \sin\theta & 0 & 0 \end{pmatrix} \begin{pmatrix} c \\ u \\ d \\ s \end{pmatrix}$$

$$\equiv \bar{\psi} C_- \psi$$

It is then easy to see that

$$[C_+, C_-] = \begin{pmatrix} 1 & 0 & 0 & 0 \\ 0 & 1 & 0 & 0 \\ 0 & 0 & -1 & 0 \\ 0 & 0 & 0 & -1 \end{pmatrix}$$

so that

$$J_0 = \tfrac{1}{2}\bar{\psi}[C_+, C_-]\psi$$

$$= \tfrac{1}{2}(\bar{c}c + \bar{u}u - \bar{d}d - \bar{s}s) \qquad (13.20)$$

which contains no $\Delta S = 1$ terms, as desired.

The unified gauge theory is saved at the price of introducing a new quark, and a new quantum number. This quantum number, if it exists, should make its effects felt in a similar way to strangeness — strong intereactions will conserve it, but weak interactions change it. (To be more precise, we see from (19) that "charged" semi-leptonic weak decays will contain $\Delta C = 0$ and $|\Delta C| = 1$ terms, whereas, from (20), "neutral" semi-leptonic decays conserve charm, just as they do strangeness, $\Delta C = \Delta S = 0$). To date, there is no direct evidence for charm; if it exists, then present accelerator energies are not high enough for charmed particles to have been produced, just as in the 1940's accelerator energies were below the threshold for strange-particle production.

But this is not all; if there are 4 quarks, then the symmetry group of the strong interactions is not SU(3), but SU(4). The hadrons, then, will group themselves into multiplets of SU(4), each multiplet containing particles of varying C. Just as a single multiplet of SU(3), for example the meson octet, contains several multiplets of SU(2) each with differing S (isodoublet with $S = 1$, isotriplet with $S = 0$, isosinglet with $S = 0$, isodoublet with $S = -1$), so the meson multiplet of SU(4) contains 15 ($=4^2 - 1$) particles.

$$\text{SU(4): 15} \begin{cases} \text{SU(3)} & \text{Charm} \\ 3^* & C = 1 \\ 8 & C = 0 \\ 1 & C = 0 \\ 3 & C = -1 \end{cases} \qquad (13.21)$$

Of these only the octet and singlet have so far been discovered. The prospect opens up a host of charmed SU(3) triplets of mesons (and of course similar multiplets of baryons) of mass >2 GeV, waiting to be discovered.

13.3 GLOBAL AND LOCAL SYMMETRIES, AND THE HIGGS MECHANISM

The unified gauge theory of weak and electromagnetic interactions has two curious features. In the first place, it is a theory which involves invariance under a symmetry group (a "weak isospin" SU(2)) but the particles which form multiplets under this symmetry group, e.g. the electron and neutrino of equation (11)

UNIFIED INTERACTIONS AND CHARM

are *not degenerate*. This is even more striking in the multiplet consisting of the charged intermediate bosons $W^\pm$ and the photon γ. To clarify the analogy with isospin, we say that there is a conserved quantum number "weak isospin" I^W, and we have the multiplets

I_3^W Mass (MeV/c^2)

$$I^W = \tfrac{1}{2} \begin{matrix} \nearrow \tfrac{1}{2}\ \nu & 0 \\ \searrow -\tfrac{1}{2}\ e^- & 0.5 \end{matrix}$$

$$I^W = 1 \begin{matrix} \nearrow 1 & W^+ & >37{,}000 \\ -0 & \gamma & 0 \\ \searrow -1 & W^- & >37{,}000 \end{matrix}$$

This lack of degeneracy (or even approximate degeneracy) is the first curious feature. The second is that the intermediate bosons *do not have zero mass*. The point here is that in a gauge theory (like electrodynamics) the spin 1 "gauge" particles must have zero mass (as the photon does). Here, the unified weak and electromagnetic theory is a gauge theory, but the W particles do not have zero mass. As mentioned above, this is due to a device called the Higgs mechanism.

Before describing this, I shall first outline what is meant by the term "gauge theory", and why such theories normally lead to particles of zero mass.

A gauge theory is a theory of "local", as distinct from "global" symmetries. Symmetries are concerned with conservation of quantum numbers, for example conservation of charge. Now if the total charge here in Canterbury suddenly decreased by 100 coulombs, and at the same time the total charge in Peking increased by 100 coulombs, then *globally* charge would be conserved, but on a *local* scale it would not. Local conservation of charge demands that a change of charge in a particular volume is accompanied by a compensating change *in the neighbouring volume*. Now let us express this mathematically. In electrodynamics conservation of charge is expressed by demanding invariance under

$$\psi(x) \to \exp(ie\lambda)\psi(x) \qquad (13.22)$$

where ψ stands for any field, and e is its charge. This means that the change (22) in the *phase* of a field leads to no new physical situation. Since λ is a constant, however, (22) seems to imply that if we change the phase of an electron in Canterbury, we should *at the same time* change the phase of an electron in Peking, or in a distant galaxy. But this is at variance with the theory of relativity, according to which no information can travel faster than light, and therefore the relative phases of two electrons in Canterbury and Peking separated by a space-like interval is *arbitary*. This implies that λ in equation (22) should not be a constant but a function of space-time $\lambda(x)$, and we have invariance under the *local* transformation

$$\psi(x) \to \exp(ie\lambda(x))\psi(x) \tag{13.23}$$

Electrodynamics is, however, only invariant under (23) if it is accompanied by the transformation

$$A_\mu(x) \to A_\mu(x) + \frac{\partial \lambda(x)}{\partial x_\mu} \tag{13.24}$$

on the 4-vector potential A_μ. This is called a gauge transformation, and its importance is that electrodynamics is only invariant under (24) if *the photon is massless*. (A mass term in the Lagrangian would have the form

$$\sum_{\mu=1}^{4} m^2 A_\mu A_\mu$$

and this is not invariant under (24) unless $m^2 = 0$). Hence we have the result that *local conservation of charge implies that the photon mass is zero*. Global charge conservation, on the other hand, has no such consequence. The step we make now is to generalise this whole philosophy to a non-abelian symmetry group, SU(2). Calling the generators of this group weak isospin, we now demand that, unlike ordinary (strong) isospin, weak isospin is *locally conserved*. The consequences of this are that

(1) weakly- and electromagnetically-interacting particles should form degenerate multiplets, differing only in their electric charge
(2) there should be 3 massless "gauge" particles of spin 1, W^+, γ, W^-, which belong to a multiplet with $I^W = 1$. (13.25)

These conclusions are the ones normally expected from an SU(2) gauge theory.

When we considered *chiral* symmetry in the last chapter, however, we saw that (see section 12.3 and table 12.2, pages 213 and 219) instead of it being realised by the occurrence of degenerate particles of opposite parity, it is realised by particles of one parity only, together with 3 massless pions (or, in the case of SU(3), 8 massless pseudoscalar mesons). This phenomenon is known as "spontaneous symmetry breakdown". We now suppose that the weak isospin symmetry is also spontaneously broken, so that, analogous to the 3 massless pions, we would expect 3 massless spin zero particles. Since, *in addition*, weak isospin symmetry is a gauge symmetry (i.e. a local symmetry), then we should expect, naively

(a) massless electron and neutrino
(b) 3 massless scalar particles, of charges +1, 0, −1.
(c) 3 massless "vector" (spin 1) particles W^+, γ, W^- (13.26)

Now it was shown by P. W. Higgs in 1964 that this does *not* happen. Applied to the particular model in hand (Weinberg 1967), what happens is that we have instead

(a) massless neutrino but massive electron
(b) 1 neutral massive scalar particle
(c) 3 vector particles, of which W^+ and W^- are massive, and γ massless (13.27)

This is, of course, what is observed in nature, apart from (b), the massive neutral scalar field, for which there are (to date) no candidates. What has happened in the passage from (26) to (27) is that the two charged scalar fields "combine" with W^+ and W^- to supply the extra degree of freedom required for these fields to become massive (for it must be remembered that massless spin 1 fields have two spin projections $S_z = 1, -1$, whereas massive fields have three $S_z = 1, 0, -1$).

In summary, we have a theory of weak and electromagnetic interactions which

(a) is a gauge theory, and is therefore renormalisable; and contains
(b) a massive electron and massless neutrino (and, similarly, a massive muon and massless muon neutrino)
(c) two charged massive intermediate bosons $W^\pm$ a massless photon γ, and a massive neutral boson Z°*
(d) a massive neutral scalar particle (the "Higgs" particle, so far unobserved)
(e) charmed hadrons, so far unobserved.

13.4 A NOTE ON THE NEW NARROW RESONANCES

In November 1974 two boson resonances were discovered at 3.1 and 3.7 GeV. The first of them, generally denoted ψ, was discovered independently by the MIT-Brookhaven group in the reaction $p + Be \to e^- + e^+ +$ anything, by discovering a peak in the invariant mass of $(e^+ e^-)$, and by the Stanford-Berkeley group in electron-positron annihilation $e^+ + e^- \to$ anything. The second resonance, ψ' was announced two weeks after ψ by the California group. The remarkable feature of these resonances is their *narrow width*. $\psi(3105)$ is too narrow for its width to show up directly on the experimental curve. The experimental width was given as $\Gamma < 1.3$ MeV, and in fact it is estimated that the width for decay into hadrons is given by $\Gamma = 50$ keV. A similar situation

---

* This extra Boson Z° occurs because the symmetry group is actually SU(2) × U(1), not simply SU(2).

holds for ψ' (3695). For purposes of comparison, it is instructive to compare the production and subsequent delay of ψ by e^+e^- annihilation

$$e^+ + e^- \to \psi \to \text{hadrons}$$

with the corresponding production and decay of ρ (see section 6.5)

$$e^+ + e^- \to \rho \to \text{hadrons}$$

first observed in 1967. The width for ρ decay into hadrons, however, is $\Gamma_\rho \sim$ 150 MeV. The ψ particles are about 10^3-10^4 times longer lived than most hadronic resonances, typified by ρ.

There are already several theories to explain the narrow width. Since understanding of the ψ particles will probably develop rapidly, it would be pointless to give here a detailed account of the present situation. I shall confine myself to some general observations which should give an idea of the range of possibilities that has been opened up. For up-to-date information, the reader is best advised to consult the latest editions of, for example, Physical Review Letters.

(a) An obvious possibility is that the ψ's are not hadrons. Since the spin of ψ (and ψ') is quite probably 1, may they be the intermediate boson? If they are, it would have to be explained why there are two of them, both neutral, and (at least so far) no charged bosons discovered. Also, according to the Weinberg-Salam theory, the mass of W is at least 37 GeV (equation (17)), so ψ cannot be W. If the ψ's are weak bosons, we should expect parity violation in the decays, which would give an up-down asymmetry, similar to that observed in Λ decay (see section 8.7)

(b) A currently popular theory is that the ψ mesons have "hidden charm", i.e. they are $(\bar{c}c)$ states. In order to explain their long lifetime, a mechanism by the name of "Zweig's rule" is invoked. According to this rule, the normal strong decay of a meson $(\bar{a}a)$ is

$$(\bar{a}a) \to (\bar{a}q) + (a\bar{q}) + \text{---}$$

So, for example, if a meson is made out of strange quarks only $(\bar{s}s)$, then the decay products will include such mesons as $(\bar{s}u), (\bar{s}d), (s\bar{u})$ or $(s\bar{d})$ — in other words, *strange* mesons. By consulting the particle lists at the beginning of the book, it is seen that the φ meson decays with a large width into $\bar{K}K$, but its decay into 3π is strongly suppressed. Also, from equation (11.34) it is seen that φ is precisely a $(\bar{s}s)$ bound state. Thus Zweig's rule operates in φ decay, explaining why the branching ratio into non-strange particles is small. If $\psi = (\bar{c}c)$ and Zweig's rule is applied, the preferential decay will be into *charmed* mesons, but if these are heavy (because, presumably c itself is heavier than s,

UNIFIED INTERACTIONS AND CHARM

d and u), then this decay will be energetically forbidden, and the width of ψ will be small. If the ψ's are indeed ($\bar{c}c$) bound states, then charmed particles should be expected at masses of about 2 GeV, as discussed in the last section.

(c) It may be that ψ particles cannot be accounted for in the context of existing theories. As an example, it may be that there is a new interaction in nature, hitherto unnoticed, which is weaker than the strong, but stronger than the electromagnetic.

13.5 CONCLUDING REMARKS

In this book I have tried to give an account of some of the progress that has been made in elementary particle physics over the last few decades, by studying their symmetries. The picture that emerges is not a simple one, and I should like to conclude with one or two comments.

To begin with, the symmetries themselves are of four types

(1) Continuous space-time symmetries
    rotations in space
    displacements in space
    displacements in time
    Lorentz transformations
(2) Discrete space-time symmetries
    space inversion
    time reversal
(3) Gauge transformations
    charge gauge
    baryon and lepton gauges
    weak isospin and hypercharge gauges
(4) Strictly "internal" symmetries (not to do with space and time)
    isospin
    unitary spin
    chiral symmetries

Symmetries of the first and third types are exact, and the conservation laws they give rise to are exact conservation laws; for instance, conservation of energy-momentum and charge. The symmetries of the first type hold because of the *properties of space-time itself,* and perhaps this is the reason that they are universal symmetries.

In contrast, the symmetries of space and time *reversal,* (2), are only approximate. Space inversion symmetry is violated by weak interactions, and time reversal by some interaction, as yet unknown; perhaps the electromagnetic interaction, perhaps a hitherto unsuspected superweak interaction, perhaps neither of these. It was a great surprise to discover, in 1957 and 1964, that these discrete symmetries are not universal.

There is one discrete symmetry which is believed, on both theoretical and experimental grounds, to be exact, and that is CPT. It is not a pure space-time symmetry, however, for C, which converts particles into antiparticles, is unconnected with space and time.

Next there are the "internal" symmetries (4) which are *all* only approximate. They may be dubbed, for convenience, symmetries of *interactions*. The leading question about them is, why are they only approximate? Or, contrariwise, why are they there at all? In this book I have made the conventional distinction between strong, weak and electromagnetic interactions, tacitly assuming that they are independent. This may not be true, however; there seem to be grounds for believing that the different interactions "know about each other". Recall that weak interactions violate parity; this requires the presence of vector and axial vector currents, of opposite parity. These currents have the quantum numbers of the isospin and strangeness selection rules. The vector currents, when commuted together, give rise to a symmetry group (SU(2) or SU(3)) which is indeed the symmetry group of the strong interactions. Moreover, as has recently been discovered, the same is true of the axial vector currents. Inclusion of them enlarges the symmetry of the strong interactions to SU(2) × SU(2) or SU(3) × SU(3). These symmetries, however, are realised in a different way from SU(2) and SU(3). Instead of degenerate multiplets, they manifest themselves by the presence of massless particles (or, since the symmetries are only approximate, of *almost* massless particles).

Thus, as well as seeing that there is a connection between the weak and strong interactions, we have also seen that there are two distinct ways in which a symmetry may be realised; either through degenerate multiplets or through massless particles.

Finally, there is the unified gauge theory of weak and electromagnetic interactions. Here the symmetry manifests itself neither by degenerate multiplets nor by the occurrence of massless spin zero particles. Indeed, it is not at all obvious that the symmetry is there at all! If it is there, however, we should expect to discover the Higgs particle; and, at least in one version of the theory, we should also expect charmed hadrons. If charm is ever discovered, then to the extent that quarks are real, the "fundamental" particles will then be 4 leptons $(e^-, \nu_e, \mu^-, \nu_\mu)$, 4 quarks (c, u, d, s), the photon γ and intermediate bosons $W^\pm, Z^0$, and perhaps some extra bosons to "glue" the quarks together to make hadrons. These go by the name of "gluons". Because of the 4 leptons and 4 quarks, there would be a sort of lepton-hadron symmetry. This indeed was one of the reasons for introducing a fourth quark (Bjørken and Glashow 1964).

Most physicists would agree that in the last few years great strides have been made, but the Final Solution still seems to elude us. It may be that totally new concepts are required in order for the current theories to fit into a unifying scheme; on the other hand, we may already have all the necessary information, and simply need someone to appreciate its significance.

It is reported that Avicenna the philosopher said to a Sufi: "What would there to be seen if there were nobody present to see it?" The Sufi answered: "What could *not* be seen, if there were a seer present to see it?"

FURTHER READING

Weinberg-Salam Unified Theory

A. Benvenuti et al "Observation of muonless neutrino-induced inealstic interactions" *Physical Review Letters* **32**, 800 (1974)

D. B. Cline, A. K. Mann and C. Rubbia "The detection of weak neutral currents" *Scientific American,* December 1974, p.108.

F. J. Hassert et al "Observation of neutrino-like interactions without muon or electron in the Gargamelle neutrino experiment" *Physics Letters* **46B**, 138 (1973)

A. Salam "Weak and electromagnetic interactions" in *Elementary Particle Physics* (N. Svartholm, ed.) Almquist and Wiksells, Stockholm (1968) p.367.

S. Weinberg "A model of leptons" *Physical Review Letters* **19**, 1264 (1967)

S. Weinberg "Recent progress in gauge theories of the weak, electromagnetic and strong interactions" *Reviews of Modern Physics* **46**, 225 (1974)

S. Weinberg "Unified theories of elementary particle interaction" *Scientific American* July 1974, p.50

Charm

J. D. Bjørken and S. L. Glashow "Elementary particles and SU(4)" *Physics Letters* **11**, 255 (1964)

M. K. Gaillard, B. W. Lee and J. L. Rosner "Search for Charm" Fermilab Reports 74/86 and 75/14 THY (Fermi National Accelerator Laboratory, Batavia, Illinois, U.S.A.)

S. L. Glashow, J. Iliopoulos and L. Maiani "Weak Interactions with lepton-hadron symmetry" *Physical Review* D **2**, 1285 (1970)

ψ *particles*

J. J. Aubert et al "Experimental observation of a heavy particle J" *Physical Review Letters* **33**, 1404 (1974)

J.-E. Augustin et al "Discovery of a narrow resonance in e^+e^- annihilation" *Physical Review Letters* **33**, 1406 (1974)

The CERN Theory Boson Workshop "The narrow peaks in $pp \to e^+e^- + X$ and e^+e^- annihilation" CERN report number TH 1964 (9th December 1974) CERN, Geneva, Switzerland.

Index

Adair and Leipuner   145
Adler, S.   208
Adler consistency condition   215, 218–9
Adler-Weisberger sum rule   220
Anderson and Neddermeyer   1
Angular momentum
    conservation of   18, 19, 21, 23
    operator   22
Antiparticles (see also Particles and Antiparticles)   14
    mass and lifetime   149
    parity   28
Antiquark   192
Approximate
    degeneracy   52, 55, 183–4
    symmetry   52, 170, 187, 198
Arrow of time   33, 168
Associated production   77
Axial vector current   173, 240
    partially conserved (PCAC)   213–6, 218–21

Barn   4
Baryon   11
    decuplet   185–6, 195, 197, 198, 201–2
    number   12, 19
    octet   182–4, 195, 197–8, 202
    resonances (spectrum)   127
    singlet   187, 197, 198
Bernstein, Feinberg and Lee   163
Beta decay
    C violation in   146, 149
    CP conservation in   146–7
    muon   173–4
    neutron   171–3
    parity violation in   136–8
BeV   2
Binding energy   43–4
Bjorken, J. D.   110, 240
Boundary conditions   36–7
Broken symmetry   187, 198, 236

C   35, 63, 156, 240
    conservation   36–7
    violation   146, 149, 163–4
C (charm) – see Charm
Cabibbo
    theory   208–13, 232
    angle   210–3, 226
Cascade particle – see Ksi
Charge
    asymmetry   165
    conjugation   35–6, 63, 163
    conservation   12, 19, 235–6
    distribution   97, 101–4, 106–7
    families   64–5, 78
    independence   45–6, 52–4
    parity   19, 35
    point   102–3
    radius   103
    symmetry   43–5
    weak   176, 179
Charm   224, 230, 231–4, 237–40
Chew – Frautschi diagram   130
Chiral symmetry   208, 213, 216–9, 236, 239
Christenson, Cronin, Fitch and Turlay   159
Clebsch – Gordon coefficients   69–73
Cobalt 60 decay   134–8
Coleman – Glashow formula   205
Colliding beams   111, 114
Commutation relations   39, 181, 216–7, 220
Commutator   20, 38
Conservation   19, 239
    of angular momentum   18, 19, 21, 23
    of baryon number   12, 19, 22
    of charge   12, 19, 22, 175, 235–6
    of charge parity   19
    of CP in beta decay   146–7
    of energy   18, 19
    of isospin   51, 52, 62–3, 179
    of lepton number   12, 19, 22, 140
    of momentum   17, 19–21
    of muon number   19

Conservation—cont.
  of parity in strong and electromagnetic
    interactions   19, 29–30
  of unitary spin   187
Conserved vector current (CVC)   175–9,
    208, 219
Conversion and regeneration of K mesons
    152–3
Cosmological field   165
Coulomb   5
  interaction   43–5
Cowan   74
CP conservation
  and K decays   156–8
  in beta decay   146–7
CP violation   159–68
  analysis of puzzle   159–63
  and arrow of time   168
  and $\Delta I = 3/2$ nonleptonic interactions
    162–3
  and electromagnetic interactions   163
  experimental results   164
  superweak theory   164–5
CPT   148, 240
  in K decays   166
  theorem   148–9
Crawford et al   145
Cross section   4
  inclusive   109
  strong interaction   9
Current
  algebra   208, 220–2
  axial vector   173, 240
  Cabibbo form   232
  conserved   22
  electromagnetic   175, 229
  isospin   179
  vector   173, 175–9, 240
Current–current interaction   172, 225

Dalitz plot   124–6, 132–4
Decay times   9
Decuplet   185–6, 188, 194–5, 207
Delta ($\Delta(1236)$)   11, 116–9, 185–6,
    201–2
$\Delta I = \frac{1}{2}$ rule – see Selection rules
$\Delta S = 1$ – see Selection rules
$\Delta S = \Delta Q$ decays   89–92, 209
Detailed balance, principle of   23, 33
Deuteron   30, 59–65, 79, 95, 128

Dirac   25, 98–9, 172
  delta function   101–2
  equation   139, 141

Eightfold way   182–4
Einstein, theory of gravity   9
Eisler et al   145
Elastic scattering   97, 110
Elementarity of particles   128–30, 206–7
Electric form factor   104–6
Electromagnetic field   6
Electromagnetic interactions   4–7, 10,
    41, 224–37
  and C conservation   36
  and C (CP) violation   163–4
  and isospin   42, 51, 94–6
  and parity   29–31
  and strangeness   94–6
  and time reversal   33–5, 163
  and U-spin   205
  selection rules   94–6
Electromagnetic mass differences   42,
    204–5
Electromagnetic structure of nucleons
    97–115
Electron   1, 5, 11, 47, 86, 149, 229–30,
    235, 237
  and neutrino   234–5, 237
Electron–electron scattering   97
Electron–positron annihilation   111–4,
    237–8
Electron–neutron scattering   97
Electron–proton scattering   97, 100–11
Equal spacing rule   186, 202
Eta (η) meson   125–6, 184, 197, 199–201
  decay   36, 95–6, 163
  η' 197, 199–200
Exchange force   55
Exotic states   128, 130

Fermi   116
  coupling constant   34, 149, 171–3,
    208–13, 221, 225–6
  golden rule   33
  transition   171–3, 176–7, 209
  weak interaction   8, 224–5
Fermi-Yang model   190
Feynman, R. P.   106, 109, 110, 170, 177
Feynman and Speisman   42
Feynman diagram   6

INDEX

Fine structure constant 99
Fitch, Roth, Russ and Vernon 166
Form factors 100–7
  electric and magnetic 104–7
  proton and neutron 100–4
Formation experiments 120
Four-vector 2
Fraser and Fulco 106
Friedman and Telegdi 143

G – parity 63–4
Gamow-Teller
  coupling constant 149, 171–3, 208, 221
  transition 149, 171–3, 210
Garwin, Lederman and Weinrich 143
Gauge particles 235
Guage transformations 19, 22, 224, 236, 239
Gell-Mann, M. 93, 170–1, 177, 180, 184, 192, 208, 217, 220
Gell-Mann and Pais 93
Gell-Mann – Nishijima relation 78–82, 92, 94, 96
Gell-Mann – Okubo mass formula 198, 204
Gell-Mann – Pais theory of kaons 150–3
Generator 20
  of chiral symmetry 220
  of isospin rotations 50
  of I-spin, U-spin and V-spin 181–2
  of rotations 22, 39, 50
  of translations 20, 38
Gershtein and Zeldovich 177
GeV 2
Glashow, Iliopoulos and Maiani 232
Gluons 240
Goldhaber $et\ al$ 141–2
Good $et\ al$ 155
Good quantum number 52, 64, 78
Gravitational interaction 9
Graviton 9
Groups 37–40, 181
  abelian 38
  of rotations 39, 40, 50, 181, 216
  of translations 38
Gürsey, F. 165

Hadrons 11
Hard core 107

Heisenberg, W. 46
Helicity 13, 138–9 ·
Hidden symmetry 216
Higgs, P. W. 237
Higgs mechanism 227, 234–7
Higgs particle 237, 240
Hypercharge (Y) 80, 184, 187
  weak 227, 239
Hyperfragments 82

Inclusive cross section 109
Inelastic scattering 97, 108, 110
Interaction ($see\ also$ Electromagnetic, Strong and Weak Interactions)
  and cross sections 4
  and decay times 9
  current–current 172, 225
  new 239
  strong, weak, electromagnetic and gravitational 6–9, 166–7, 240
  superweak 164–5
Intermediate boson (W, of weak interactions) 8–9, 224–5, 227, 229, 231, 235–8, 240
Internal conversion 83
Invariance 19
  CPT 148
  isospin 51
  scale 110
  space inversion 25–6
  time reversal 149
  under rotations 18, 19, 22
  under spatial translation 17, 19, 20
  under time displacement 18, 19
  unitary spin 184
Isospin 40, 41–65, 170, 183, 188, 190–2, 239
  addition 58, 66
  conservation 51–2, 62–3, 78, 82, 179
  current 179
  of strange particles 78–82
  multiplets 191–2
  nucleon 46–7
  pion 54–7
  selection rules 85–96, 158–9
  transformation 47–51
  wave function 60, 61
  weak 227, 234, 239

Isotopic spin − see Isospin

K − see Kaon
K* 11, 126−8, 129, 197, 202, 204
Kaon (K) 11, 75, 80, 135, 150, 184, 199−201
  conversion and regeneration 152−3
  CP invariance in decay 156
  CP violation in decay 159−68
  decay 33, 90−4, 135−6, 148−69, 230
  discovery 75−6
  Gell-Mann−Pais theory of 150−3
  interferometry 153−5
  isospin 81
  K_1 and K_2 151, 153−6
  lifetimes 151
  mass 80
  parity 82−3
  strangeness 77−8
Kemmer, N. 54
Ksi (Ξ) 76, 170, 182−4, 202, 205
  decay 76−8, 91−2
  discovery 76
  isospin 81−2
  Ξ (1530) 121, 185−6, 201−2
  mass 80
  parity 83
  spin 145−6
  strangeness 78

Lagrange's equations 17
Lambda (Λ) 11, 75, 170, 182−4, 190−1, 202
  decay 75, 89, 92, 143−5, 149
  discovery 75
  isospin 79−80
  Λ (1405) and Λ (1520) 120−1
  mass 80
  spin 145−6
  strangeness 77−8
Landau, L. 139
Lee, T. D. 36, 163
Lee and Yang 134−5, 139, 145, 158
Legendre polynomial 24, 118
Lepton 11, 240
  -hadron symmetry 240
  number 12, 19, 140
Leptonic weak interactions 86−92
Lightlike momentum 101

Lüders, G. 148

Magnetic form factor 104−6
Magnetic moment 98, 149
  anomalous 99, 100
Mass formulae 198−205
Massless particles (see also Pion, massless) 240
Matter and antimatter, distinction between 165, 166−8
Medium strong interaction 188
Meson (see also Eta, Kaon, Muon, Phi, Pion, Psi, Rho, Omega) 1, 11
  octets and singlets 195−7
  resonances (spectrum) 129
MeV 1
Michel, L. 64
Mirror nuclei 43−6
Mixing
  angle 203−4
  of $K°$ and $\bar{K}°$ 150, 155, 164
Momentum
  conservation of 17, 19, 20, 21
Mott scattering 104−5, 110
Muon 1, 5, 11, 12, 86, 149, 188, 137
  number 13
  decay 86, 143, 173−4
  neutrino 2, 3, 11, 237

Nambu, Y. 77, 214, 216
Ne'eman, Y. 171, 187
Neutral currents 224, 230−2
Neutrino 1, 4, 5, 11, 86, 229−30, 235, 237, 240
  and electron 234−5, 237
  and parity violation 138−41
  electron 4, 234−5, 237
  existence of two 13
  helicity 13, 139−42
  mass 140, 235, 237
  muon 2, 3, 11, 237
  two component 139
Neutron (see also Nucleon) 1, 11, 41−3, 46−51, 64−5, 170, 183, 187, 190−1, 205, 206
  decay 8, 9, 13, 86, 171−3
  dipole moment 34−5, 164
  form factors 104−7
  hard core 107

INDEX

Newton
  and gravity   6, 18
  law of motion   17
Nishijima, K.   77
Nonleptonic weak interactions   86, 92–4, 143–5
Nuclear
  democracy   130, 207
  forces   43–6
  interaction   7
  magneton   99, 104
Nucleon   46, 79, 170, 183, 190, 202, 205
  -antinucleon bound state   56
  electromagnetic structure of   97–115
  electrons and   97–100
  hard core   107
  isospin   46–7, 66–9
  point-like constituents of   110

Octet   182–4, 188, 194–8, 202
Okun' and Pontecorvo   156
Omega (ω) meson   11, 107, 112–3, 122–5, 129, 197, 203–4
Omega-phi mixing   203–4
Omega minus (Ω^-)   11, 121, 186, 202, 207
Oneda, S.   77
Optically active compounds   36
Orbital excitation   196–7, 206

P – see Parity
Pais, A.   25, 77, 93, 150–3, 165
Pais and Piccioni   152
Parity   24–8
  and angular momentum   26
  conservation   19, 29–30
  doublet theory   133
  doubling   213
  impurity   29
  intrinsic   27
  isotopic   64
  of particle and antiparticle   28
  of photon   30
  of pion   30, 31
  relative   28, 83–4
  violation   19, 25, 36, 132–47, 239, 240
Particles and antiparticles   36, 37, 54, 80
  distinction between   19
Partons   110

Pauli, W.   148
Pauli exclusion principle   31, 48, 59
  generalised   59, 93–4, 158
Pauli spin matrices   48
Phase shift analysis   119
Phi (ϕ) meson   107, 112–3, 125, 129, 197, 202–4
  decay and Zweig's rule   238
Photon   1, 4, 5, 6, 9, 11, 94, 224, 235, 236, 237, 240
  parity   30
  timelike   111–4
  virtual   6, 99, 101
Piccioni   152, 155
Pion   1, 7, 11, 52–4, 64–5, 74, 79, 184, 190, 199–201
  clouds   97–8, 104, 175
  decay   7, 12, 13, 88, 94, 143, 176–7, 214
  G-parity   63–4
  isospin   54–7, 66–9
  mass   2, 214
  massless   215, 222, 236
  parity   30, 31
  soft   215
  spin   23, 24
  virtual   98
Pion-nucleon system
  isospin   67–9
  resonance   116
  scattering   116, 220–2
Point
  charge   97, 102–3
  -like constituents of nucleon   110
  particle   98, 104–5
Powell, C.   1
Production experiments   120
Proton (see also Nucleon)   1, 5, 11, 12, 41–3, 46–51, 64–5, 170, 183, 187, 190–1, 205, 206
  form factors   104–7
  hard core   107
Pseudoscalar   31
  mesons   184, 195, 199–201
Psi (ψ) particles   237–8

Quantum electrodynamics   99, 224
Quantum field theory   6
Quantum mechanics   4, 10
Quantum theory   59, 153

Quark 110, 192–6, 206, 224, 230, 232–4, 240
  existence 206
  fourth (charmed) 224, 230, 232–4, 240
  model 195–8, 217, 226

Radiative transition 85
Raising and lowering operators 48–9, 56–7, 66 *et seq*, 179 *et seq*
Ramsey 34
Regge
  poles 8, 116, 130
  trajectories 130
Renormalisable interactions 224, 227
Resonances 10, 116–131
  baryon (spectrum) 127
  compositeness of 128–30
  exotic 128–30
  meson (spectrum) 129
Rho (ρ) meson 11, 107, 112–3, 122, 129, 202, 204, 238
Right and left
  asymmetry 135
  distinguishability and indistinguishability 136, 138
Rochester and Butler 74
Rosenbluth formula 104
Rotation 49
  group 39, 40, 50, 181, 216

Sakata model 170, 184, 190–1
Salam, A. 139, 224
Scale invariance 110
Scattering
  elastic 97, 110
  electron–electron 97
  electron–neutron 97
  electron–proton 97, 100–11
  inelastic 97, 108, 110
  pion-nucleon 116, 220–2
Schrödinger equation 18, 25, 32, 130
Schwinger, J. 148
Selection rules 85–96, 180, 240
  $\Delta S = 1$ 78, 155, 230, 234
  $\Delta S = 0, \Delta I = 1$ 89, 95, 178, 180, 212–3
  $\Delta S = 1, \Delta I = \frac{1}{2}$ 92, 93, 180, 212–3
  $\Delta S = \Delta Q$ 89–92
  $\Delta I = 1/2$ in K decay 158–9, 162–3

Selection rules–cont.
  $\Delta I = 3/2$ 90, 92, 162
  and unitary symmetry 178–86
Semi-leptonic weak interactions 86–92
Sextet 194
Sigma (Σ) 11, 75, 170, 182–4, 202, 204
  decay 7, 89, 90, 92, 94
  discovery 75–6
  isospin 79–80
  mass 80
  parity 83
  Σ (1385) 119–20, 185–6, 201–2
  spin 145–6
  strangeness 77–8
SO_3 40, 50, 181
Space inversion 23, 24–6, 32, 64, 239
Spacelike momentum 101
Spin 11, 23, 24, 47
Spin wave function 59, 60
Storage rings 111
Strange particles
  discovery of 74–6
  isospin of 78–82
Strangeness 74–84, 180, 183, 188, 190
  quantum number 77–8
  selection rules 78, 85–96
Strong interactions 4, 7–10, 41
  and C conservation 36
  and chiral symmetry 208
  and isospin 51, 52, 78
  and parity 29–30
  and strangeness 77
  and time reversal 33–5
  and unitary symmetry 184–8
  cross section 9
Structure factors 109–111
SU_2 40, 50, 181, 191, 213, 216, 240
$SU_2 \times SU_2$ 217–9, 220, 221, 222, 240
SU_3 (*see also* Unitary symmetry) 182, 187–8, 190–1, 192–5, 198, 207, 209, 211–3, 218–9, 240
$SU_3 \times SU_3$ 218, 219, 222, 240
SU_4 234
Supermultiplets 182–7, 192–5, 207
Superposition principle 151, 153
Superweak interaction 164–5, 239
Symmetry 19, 37
  and indistinguishability 15, 51, 136
  approximate 52
  broken 187, 198, 236

Symmetry—cont.
  charge   43–5
  chiral   208, 213, 218, 236, 239
  discrete   239
  exact and approximate   19, 22–3, 52, 239
  group   227, 240
  hidden   216
  in classical physics   15
  in quantum physics   18
  isospin   23, 47 et seq
  lepton–hadron   240
  local and global   234–7
  of strong and electromagnetic interactions   18, 23
  space inversion   23, 239
  space-time   239
  spontaneous breaking of   236
  time reversal   23, 239
  unitary   23, 170, 179–205

T – see Time reversal
Taylor, J. C.   213
Theta–tau puzzle   132–5, 150, 158, 160
't Hooft, G.   227
Timelike
  momentum   101
  photons   110
Time reversal   31–3, 148
  in electromagnetic interactions   33–5, 163, 239
  in K decays   166
  in strong interactions   33–5
  invariance in beta decay   149
  invariance in Λ decay   149
  no conserved quantum number   32
  violation   239
Transitions
  allowed   171
  Fermi   171–3, 176–7, 209
  Gamow-Teller   149, 171–3, 210

Uncertainty principle   6
Unified gauge theory   232, 234–40
Unified theory of weak and electromagnetic interactions   224–40
Unitary spin (see also Unitary symmetry)   183–4, 187, 192–3, 203–4, 239

Unitary symmetry   121, 170, 179–205, 219
  supermultiplets   182–7, 207
Units   1, 231
Universality
  Cabibbo   209–13
  in weak interactions   174
Up–down asymmetry   145
U-spin   182, 199, 200, 205

V – A theory   173
Vector addition rule   55, 65, 87, 191
Vector current   173, 179, 240
  conserved   175–9, 208, 219
Vector meson   112, 122, 202–4
V-spin   181, 199, 211

W boson – see Intermediate boson
Weak charge   176, 179, 225
Weak hypercharge   227, 239
Weak interactions   4, 5, 8–9, 10, 224–37
  and isospin   51, 85–94
  and parity violation   25, 132–47, 240
  and strangeness   77, 85–94
  leptonic and semi-leptonic   85–92, 209
  nonleptonic   85–6, 92–4, 143–5
  renormalisable   224, 227
Weak isospin   227, 234, 239
Weak neutral currents   224, 230–2
Weber, J.   9
Weinberg, S.   224, 227
Weinberg–Salam theory   224, 227–31, 238
Weyl equation   141
Width (decay)   10, 237–9
Wigner, E. P.   37, 47
Wolfenstein, L.   164–5
Wu, C. S.   134–6

Y – see Hypercharge
Y* (1385)   119, 185–6, 201–2
Yamaguchi   77
Yang, C. N. (see also Lee and Yang)   62
Yukawa, H.   1, 7, 55, 74, 102–6

Z° boson   229, 237, 240
Zeeman effect   47
Zweig, G.   192
Zweig's rule   238